NUMBER THEORY

PLOWING AND STARRING THROUGH HIGH WAVE FORMS

Series on Number Theory and Its Applications* ISSN 1793-3161

Published

Vol. 4 Problems and Solutions in Real Analysis
by Masayoshi Hata

Vol. 5 Algebraic Geometry and Its Applications
edited by J. Chaumine, J. Hirschfeld & R. Rolland

Vol. 6 Number Theory: Dreaming in Dreams
edited by T. Aoki, S. Kanemitsu & J.-Y. Liu

Vol. 7 Geometry and Analysis of Automorphic Forms of Several Variables
Proceedings of the International Symposium in Honor of Takayuki Oda on the Occasion of His 60th Birthday
edited by Yoshinori Hamahata, Takashi Ichikawa, Atsushi Murase & Takashi Sugano

Vol. 8 Number Theory: Arithmetic in Shangri-La
Proceedings of the 6th China–Japan Seminar
edited by S. Kanemitsu, H.-Z. Li & J.-Y. Liu

Vol. 9 Neurons: A Mathematical Ignition
by Masayoshi Hata

Vol. 10 Contributions to the Theory of Zeta-Functions: The Modular Relation Supremacy
by S. Kanemitsu & H. Tsukada

Vol. 11 Number Theory: Plowing and Starring Through High Wave Forms
Proceedings of the 7th China–Japan Seminar
edited by Masanobu Kaneko, Shigeru Kanemitsu & Jianya Liu

*For the complete list of the published titles in this series, please visit:
www.worldscientific.com/series/sntia

Series on Number Theory and Its Applications Vol. 11

NUMBER THEORY
PLOWING AND STARRING THROUGH HIGH WAVE FORMS

Proceedings of the 7th China–Japan Seminar

Fukuoka, Japan 28 October – 1 November 2013

Editors

Masanobu Kaneko
Kyushu University, Japan

Shigeru Kanemitsu
Kinki University, Japan

Jianya Liu
Shandong University, China

NEW JERSEY • LONDON • SINGAPORE • BEIJING • SHANGHAI • HONG KONG • TAIPEI • CHENNAI

Published by

World Scientific Publishing Co. Pte. Ltd.

5 Toh Tuck Link, Singapore 596224

USA office: 27 Warren Street, Suite 401-402, Hackensack, NJ 07601

UK office: 57 Shelton Street, Covent Garden, London WC2H 9HE

Library of Congress Cataloging-in-Publication Data

China-Japan Seminar on Number Theory (7th : 2013 : Fukuoka-ken, Japan)

Number theory : plowing and starring through high wave forms : proceedings of the 7th China-Japan seminar, Fukuoka, Japan 28 October–1 November 2013 / edited by Masanobu Kaneko (Kyushu University, Japan), Shigeru Kanemitsu (Kinki University, Japan), Jianya Liu (Shandong University, China).

pages cm. -- (Series on number theory and its applications ; volume 11)

Includes bibliographical references and index.

ISBN 978-9814644921 (hardcover : alk. paper)

1. Number theory--Congresses. I. Kaneko, Masanobu, editor. II. Kanemitsu, Shigeru, editor. III. Liu, Jianya (Mathematics professor) editor. IV. Title.

QA241.C645 2013

512.7--dc23

2014045569

British Library Cataloguing-in-Publication Data

A catalogue record for this book is available from the British Library.

Printed in Singapore

PREFACE

The present volume is the proceedings of the 7th China–Japan Seminar on Number Theory "Plowing and Starring Through High Wave Forms" held during October 28–November 1, 2013 at the Ito Guest House, Kyushu University Moto-oka, Fukuoka, Japan; the organizers being Shigeru Kanemitsu and Jianya Liu with Professor Masanobu Kaneko as the local organizer.

The seminar was held during October 28–November 1 in Fukuoka Prefecture, but in the West of Fukuoka City, which was known as Ito-koku (Ito kingdom) in ancient times (while the 2nd seminar was held in Iizuka, Fukuoka Prefecture in 2001). There used to be more free traffic flow of people and ships between Japan and China through Korea and the braves used to sail the sea between, thus, plowing through high waves and we naturally associated forms as well. Starring is reminiscent of Starry Waves (K. O'Brien), Starry Night (van Gogh — as they sailed at night), etc.

Now about the contents of the seminar and the present proceedings. The talks ranged over a wide spectrum of contemporary number theory and related topics. As can be seen from the program at the end of this preface as well as in the following brief descriptions, in this volume, we assemble far-reaching survey papers both in Analytic Number Theory (Universality by KMa, Modular relation by ACK), Algebraic Number Theory (Complex multiplication by KMi, Distribution of roots of a polynomial by YK), and Collection of problems in number theory (SZW, CTX).

Here is a brief description:

T. Arai, K. Chakraborty and S. Kanemitsu's "On modular relations" is, an extract from, and supplement to, the coming book of the second editor and Professor H. Tsukada, "Modular Relation Supremacy" and assembles those results which were either not touched on or sketchily touched, with further elucidations. By a modular relation we mean any equivalent to, or consequences of, the functional equation satisfied by the zeta-function in question. In the paper, we concentrate on a few selected subjects including the resurrection of Koshlyakov's unduly forgotten paper in which he treats the Dedekind zeta-function of a number field with degree $\varkappa \leq 2$, i.e. rational field and the quadratic fields. We take a viewpoint of A. A. Walfisz [AAW1], [AAW2] to abuse the functional equation to mean two objects. For example, the Dedekind zeta-function for an imaginary quadratic field also means one for a positive definite quadratic form and the Dedekind zeta-function for a real quadratic field also covers the one for indefinite quadratic form and the Dirichlet divisor problem. This standpoint is adopted by H. Cohen, too. We shall also state the general $\varkappa$ case. We shall then state another class of modular relations, i.e. integrated ones, which includes classical equality due to Ramanujan et al. The method is integration in the parameter and this naturally leads us to the Riesz sums. We state some recent results of Katsurada in terms of the modular relation. We shall also streamline some papers which makes careless use of the first Riesz sum. Unlike the Hecke gamma transform, this cannot be applied as it stands. One has to integrate it at least twice as in A. A. Walfisz [AAW1], [AAW2] or apply the truncated form.

N. S. Koshlyakov, Investigation of some questions of analytic theory of the rational and quadratic fields, I-III (Russian), *Izv. Akad. Nauk SSSR, Ser. Mat.* **18** (1954), 113–144, 213–260, 307–326 with Errata.
H. Cohen, q-identities for Maass waveforms, *Invent. Math.* **91**(1998), 409–422.
[AAW1] A. A. Walfisz, On sums of coefficients of some Dirichlet series, *Soobšč. Akad. Nauk Grundz. SSR* **26** (1961), 9–16.
[AAW2] A. A. Walfisz, On the theory of a class of Dirichlet series, *ibid.* **27** (1961), 9–16.

T.-X. Cai, Y. Zhang, and Z.-Y. Shen's "Figurate primes and Hilbert's 8th problem" is centered around many classical conjectures revisited from the viewpoint of figurate primes, where a figurate prime is defined as the binomial coefficient

$$\binom{p^a}{i},$$

where $a \geq 1$, $i \geq 1$ and p is a prime in association with figurate numbers. The Diophantine equation

$$\binom{p^a}{i} - \binom{q^b}{j} = k, \tag{0.1}$$

is studied, where p, q are primes, $a, b, i, j, k \in \mathbb{N}$. Many special cases give rise to elliptic curve whose integer points are mostly found by **Magma** program. This provides a good opportunity for us to look at classical conjectures anew.

In Y. Kitaoka's "Statistical distribution of roots of a polynomial modulo prime powers", the author continues his unique research on the distribution of roots of a polynomial modulo a prime, and in the present paper, it is raised to (modulo) prime powers p^m, where one of the parameters is fixed and the other goes to infinity. As with many of his outstanding achievements, the author starts from a keen observation on a phenomena which many would have let pass by, and this research is not an exception and has started from a simple observation on the decimal expansion of rationals which is found to be related to the simple polynomial $x^n - a^n$. The second editor is proud that the author's first observations appeared in the author's book "Introduction to Algebra" published by the first editor's publishing house. This naïve observation is sublimated to a new look at the distribution of roots of a polynomial modulo prime powers.

Throughout, the polynomial $f(x) = x^n + a_{n-1} + \cdots + a_0 \in \mathbb{Z}[x]$ is fixed once and for all unless otherwise specified. There are many densities defined which are all with regard to the density of the set of primes

$$Spl(f) = \{p \mid f(x) \text{ is completely splitting over } \mathbb{Q}_p\}.$$

For $p \in Spl(f)$, $f(x)$ is decomposed as

$$f(x) \equiv \prod_{i=1}^{n} (x - r_{i,p^m}) \bmod p^m, \tag{2}$$

where among the rational integer solutions r_{i,p^m} the ordering

$$0 \leq r_{1,p^m} \leq \cdots \leq r_{n,p^m} \leq p^m - 1 \tag{3}$$

is assumed. It is easily seen that there is an integer k_{p^m} such that

$$k_{p^m} p^m := a_{n-1} + \sum_i r_{i,p^m}. \tag{5}$$

This quantity $k_{p^m} = k_{p^m}(f)$ and generalization thereof plays an important role in the paper. For example, in §2.2, the following generalization of k_{p^m}

$$R_f := a_{n-1} + \sum_{i=1}^{n} R_i \tag{10}$$

is introduced.

In the paper, the author uses the new term "decomposable" to mean that the polynomial is a composite function of two polynomials. This case along with reducible case is to be excluded.

Expectation 1 states for irreducible indecomposable polynomials f: For a natural number k with $1 \leq k \leq n-1$ The limit $Pr(f)[k] = \lim_{X\to\infty} Pr_X(f)[k]$ exists and $Pr(f)[k] = E_n(k) := A(n-1,k)/(n-1)!$, where

$$Pr_X(f)[k] := \frac{\#\{p \in Spl_X(f) \mid k_{p^m} = k\}}{\#Spl_X(f)}$$

and where $A(n,k)$ $(1 \leq k \leq n)$ are Eulerian numbers.

The density

$$Pr(f,m,L,\{R_i\}) := \lim_{X\to\infty} \frac{\#\{p \in Spl_X(f) \mid (2),(3),(8)\}}{\#Spl_X(f)} \tag{9}$$

is introduced, where

$$r_{i,\,p^m} \equiv R_i \bmod L \quad (i = 1,\dots,n). \tag{8}$$

Conjecture 1 states: If a polynomial $f(x)$ is irreducible and indecomposable,

$$Pr(f,m,L,\{R_i\}) = pr(f,m,L,\{R_i\}) \text{ if } \deg f \leq 5,$$

where pr is defined by

$$pr(f,m,L,\{R_i\}) := \sum_{K,\,q} E_n(K)\cdot\frac{1}{L^{n-1}}\cdot\frac{1}{[\mathbb{Q}(\zeta_L):\mathbb{Q}(f)\cap\mathbb{Q}(\zeta_{L/d})]}, \tag{10}$$

where $q \in (\mathbb{Z}/L\mathbb{Z})^\times$ and K satisfy the conditions

$$\begin{cases} [[q]] = [[1]] \text{ on } \mathbb{Q}(f)\cap\mathbb{Q}(\zeta_{L/d}), \\ R_f \equiv Kq^m \bmod L, \\ 1 \leq K \leq n-1, (K,L) = d, \end{cases}$$

which motivate the definition (10).

It is reported that the irreducible indecomposable polynomial $f_1 = x^6 - 2x^5 + 18x^4 - 22x^3 + 163x^2 - 116x + 631$ violates the previous conjectures and the author has been trying to rule out the exceptional cases in addition to reducible and decomposable ones.

K. Matsumoto's "A survey on the theory of universality for zeta and L-functions" is a timely far-reaching survey by the first-person singular of the theory. Timely because as the author divides the period into three, everything started about 40 years ago when S. M. Voronin proved his universality theorem to the effect that any non-vanishing holomorphic function can be approximated uniformly by a certain shift of $\zeta(s)$, the Riemann zeta-function. "S. M. Voronin, Theorem on the "universality" of the Riemann zeta-function, *Izv. Akad. Nauk SSSR Ser. Mat.* **39** (1975), 475–486".

The first period of 1975–1987 is an important one because many of the most important aspects of the theory of universality appeared during this time.

In the period 1987–1996, there were much less publications and is regarded as a rather unfertile period which was waiting for its flourishing in the coming years.

The second period: 1996–2007. The period starts with the publication of A. A. Laurinčikas' first book on universality, "A. Laurinčikas, *Limit Theorems for the Riemann Zeta-function*, Kluwer, 1996", and this period was quite rich and many important papers were published.

It ends in 2007, when J. Steuding published his Lecture Notes, "J. Steuding, Value-distribution of L-functions, Lecture Notes in Math. **1877**, Springer, 2007", and the third period started from this very year of 2007 and continues to the present.

In the pre-history section, the author mentions earlier work of Voronin's multi-dimensional analogue of the theorem of Bohr and Courant and subsequent Bohr-Jessen paper(s). It reminds the second editor when the author started his research on value distribution of zeta-functions. He read the Bohr-Jessen paper and initiated his own contributions, "H. Bohr and B. Jessen, Über die Werteverteilung der Riemannschen Zetafunktion, Erste Mitteilung, Acta Math. **54** (1930), 1–35; Zweite Mitteilung, ibid. **58** (1932), 1–55".

The modern statement of Voronin's theorem reads

Theorem 0.1. (Voronin's universality theorem) *Let K be a compact subset of the strip $1/2 < \sigma < 1$ with connected complement, and let f be a continuous, non-vanishing function on K, holomorphic in the interior of K. Then, for any $\varepsilon > 0$,*

$$\liminf_{T\to\infty} \frac{1}{T}\operatorname{meas}\left\{\tau \in [0,T] \;\middle|\; \sup_{s\in K} |\zeta(s+i\tau) - f(s)| < \varepsilon\right\} > 0 \tag{1.2}$$

holds.

Three proofs of this theorem are due, respectively, to Voronin, Good and Bagchi. The author makes a deep observation of most of the results obtained in the last 40 years, especially in the third period with his careful and thorough insights. It is hoped that as with another survey on mean square theory, this paper will be a landmark on the theory of universality in the coming decade, "K. Matsumoto, Recent developments in the mean square theory of the Riemann zeta and other zeta-functions, *Number Theory*, eds. by R. P. Bambah *et al.* Hindustan Books Agency, 2000, 241–286".

K. Miyake's "Complex multiplication in the sense of Abel" is like a picture scroll depicting historical events connected to complex multiplication between the times of Abel and the present. It is a very readable account starting from its originator's own papers and ending with the modern viewpoint, giving a glimpse and a prelude to bulkier volumes by G. Shimura or C. Birkenhake and H. Lange, etc.

The paper is written in a classic style of quoting the formulas by themselves. For example, the integral

$$z = z(w) = \int_{\gamma(w)} \frac{dw}{\sqrt{(1-w^2)(1+\mu w^2)}}$$

or the $\mathbb{Z}$-module $\mathfrak{m} = m\varpi_1 + n\varpi_2$ appears several times. Also there is no numbering of theorems or propositions and looks a little different from other papers, but to appreciate the flavor of old times in which the author took great pains to make conspicuous, we keep the style as it stands.

The epic starts from complex multiplication of elliptic functions: An elliptic function $x = f(z)$ has complex multiplication by $\alpha \in \mathbb{C} - \mathbb{R}$ if x and $y = y(z) = f(\alpha z)$ have an algebraic relation $P(x, y) = 0$ for a polynomial $P = P(x, y)$ with complex coefficients.

Then the module $\mathfrak{m} = \mathbb{Z}\varpi_1 + \mathbb{Z}\varpi_2$ of periods of a non-constant elliptic function $f(z)$ is introduced which is a $\mathbb{Z}$-submodule of $\mathbb{C}$ of rank 2 and $\tau = \varpi_1/\varpi_2$ is in the upper half-plane $\mathfrak{H}$. It expresses the above complex multiplication property by the statement that $\mathfrak{m} \cap \alpha\mathfrak{m}$ is of finite index in $\mathfrak{m}$. The field of all meromorphic functions on the complex torus $\mathbb{C}/\mathfrak{m}$ is identified with the field $\mathcal{K}_\mathfrak{m}$ of those elliptic functions which admit $\mathfrak{m}$ as their periods. The embedding Φ defined by the Weierstarss function and its derivative induces an isomorphism of $\mathbb{C}/\mathfrak{m}$ onto an elliptic curve $E_\mathfrak{m}$ over $\mathbb{C}$. The endomorphism algebra $\mathrm{End}_\mathbb{Q}(E_\mathfrak{m})$ is isomorphic either to $\mathbb{Q}$ or $\mathbb{Q}(\tau)$, the latter case holding if and only if $\mathbb{Q}(\tau)$ is an imaginary quadratic number field. Furthermore, the above finite index property of $\mathfrak{m} \cap \alpha\mathfrak{m}$ holds if and only if α is such an imaginary quadratic number that $\mathbb{Q}(\alpha) = \mathbb{Q}(\tau)$. Thus, the definition:

An elliptic curve E has complex multiplication if $\mathrm{End}_\mathbb{Q}(E)$ is isomorphic to an imaginary quadratic field.

The track from here to complex multiplication on Abelian varieties is by way of the Jacobian varieties. Let $P(X)$ be a polynomial of degree $2g+2$ with g a natural number, without multiple roots. Then the Riemann surface $\mathcal{R}_g$ of the differential form $dw/\sqrt{P(w)}$ is a hyperelliptic curve E_g of genus g defined by $E_g : y^2 = P(x)$. The vector space $\mathrm{H}^{(1.0)}(\mathcal{R}_g)$ of all holomorphic 1-forms on $\mathcal{R}_g$ is spanned by g forms $w^j\, dw/\sqrt{P(w)}, 0 \le j \le g-1$, and its dual space $\hat{\mathrm{H}}^{(1,0)}(\mathcal{R}_g)$ is taken. By the correspondence of the paths, there arises a $\mathbb{Z}$-module Π_{2g} of rank $2g$ in $\hat{\mathrm{H}}^{(1,0)}(\mathcal{R}_g) \cong \mathbb{C}^g$. Thus the complex torus $\hat{\mathrm{H}}^{(1,0)}(\mathcal{R}_g)/\Pi_{2g}$ is called the Jacobian variety $\mathrm{Jac}(\mathcal{R}_g)$ of the hyperelliptic curve.

Let Π be a $\mathbb{Z}$-module of rank $2g$ in $\mathbb{C}^g$ such that $\mathbb{R}\Pi = \mathbb{C}^g$. The complex torus $\mathbb{C}^g/\Pi$ is an Abelian variety if and only if the period matrix for Π has an alternating matrix for which the Riemann relations hold true. *A fortiori*, the Jacobian variety $\mathrm{Jac}(\mathcal{R}_g)$ is an Abelian variety.

An Abelian variety A of dimension g has complex multiplication if its endomorphism algebra $\mathrm{End}_{\mathbb{Q}}(A)$ is isomorphic to a CM field of degree $2g$, i.e. a totally imaginary quadratic extension of a totally real number field of degree g.

In this way, one can appreciate the fine views of ascending, starting from the module of periods reaching the summit of CM fields.

In Remark 2 at the end, the author mentions one of his earliest work which is a basis of the present paper, and which was published in 1971, over 45 years ago. We feel awed and at the same time we feel the importance of sustainability.

In Z.-W. Sun's "Problems on combinatorial properties of primes", the author states many conjectures on the prime-counting function $\pi(x)$, the nth prime p_n and the partition function $p(n)$ of combinatorial nature (in contrast to the author's previous article in the proceeding of the 6th seminar, "Conjectures involving arithmetical sequences"). The combinatorial properties are conjectured among exact values of these functions. §2 gives conjectures on $\pi(x)$, §3 on p_n and §4 gives those of primes related to $p(n)$. These conjectures are supported by numerical data.

The papers by Cai *et al.*, Kitaoka and Sun make rather heavy use of computers and we feel we are truly living in the digital world.

Finally, vote of thanks is due. Our seminar was supported by the JSPS (Japan Society for the Promotion of Science) and the NSFC (National Science Foundation of China). We would like to express our hearty thanks to these organizations for their generosity. We would like to thank Kyushu University for its generous permission of using its excellent facilities, especially Ito Guest House. The conference room was equipped with modern conveniences and was very useful in conducting the seminar. Thanks are also due to the devoted help of graduate students at Kyushu University, Messrs Tomoya Kiyuna, Tomohide Nakaya, Ryotaro Osanai and Koji Tasaka.

As in the case of the last few proceedings, Professor Ma Jing from Ji Lin University who assisted in its editing and we record here our hearty thanks to her for her excellent and beautiful preparation of the manuscript of the proceedings. Also we would like to thank Dr. J. Mehta from HRI for his help toward the completion of the proceedings.

As usual we complete the preface by a poem. This time Professor Jianya Liu composed it.

五十抒懷
劉建亞

甲午中秋，張雪明先生有七律見賜；金聲玉振，豪氣干雲。余以狗尾續貂作答，更兼五十抒懷。今遵金光滋先生雅屬，以之充集敘，奉求方家哂教。

少年青澀風雲會，情耽八法愛九章。
酒欲三觚酬意氣，詩須半卷寫清狂。
敢將華髮誇新雪，不信當朝無慶郎。
折桂蟾宮憑流寓，且依皓月望督亢。

注：
督亢：古燕國膏腴之地，含今固安、新城、涿州一帶。亢，古音平讀若剛。

Cover picture:
The cover picture is a modified version of a famous lithograph of *Hokusai's* "Thirty Six Views of Mount Fuji", from Kajikazawa. The lines are for analyzing the appealing image in terms of the incorporated ratios by R. Yanagi.

A legend
The compelling images — the centripetal force of Hokusai's composition. Zelazny, p. 216, l.6. "Mt. Fuji from Kajikazawa" begins with the passage:

Misted, mystic Fuji over water. Air that comes clean to my nostrils. There is even a fisherman almost where he should be, his pose less dramatic than the original, his garments more modern, above the infinite Fourier series of waves advancing upon the shore.

R. Zelazny, 24 views of Mt. Fuji, by Hokusai, in "*Frost and Fire*, Avon Books, New York 1990".

PROGRAM

Monday, October 28, 2013

13:00-13:30 Registration, opening and photo session

Afternoon Session

Monday, October 28, 13:50-14:30 Professor L. Weng (Kyushu Univ.) Motivic Euler product and its applications to Cohen-Lenstra heuristics for relative Shafarevich-Tate groups

Monday, October 28, 14:50-15:30 Professor Shigeki Akiyama (Tsukuba Univ.) On the mean divisibility of multinomial sequences

Monday, October 28, 15:45-16:00 Mr. Tomokazu Onozuka (Nagoya Univ.) Mean values of Mordell-Tornheim double zeta-functions (a joint work with Takuya Okamoto)

Tuesday, October 29, 2013

Morning Session

Tuesday, October 29, 10:00-10:40 Professor Yoshiyuki Kitaoka (Meijo Univ.) The distribution of roots of a polynomial modulo prime powers

Tuesday, October 29, 11:00-11:40 Professor G. S. Lv (Shandong Univ.) On Fourier coefficients of automorphic forms

Tuesday, October 29, 11:40-12:20 Professor Tianxin Cai (Zhejiang Univ.) Congruent numbers on the right trapezoid

Afternoon Session

Tuesday, October 29, 13:30-14:10 Professor Ryotaro Okazaki (Doshisha Univ.) Geometry for totally imaginary quartic Thue equation

Tuesday, October 29, 14:30-15:10 Professor W. G. Zhai (China Univ. of Mining and Tech.) On the Dirichlet divisor problem in short intervals

Tuesday, October 29, 15:30-15:45 Mr. Tomoya Kiyuna (Kyushu Univ.) A certain differential equation for Jacobi forms

Tuesday, October 29, 15:45-16:00 Mr. Takahiro Wakasa (Nogoya Univ.) The explicit upper bound of the multiple integral of $S(t)$ on the Riemann Hypothesis

Wednesday, October 30, 2013

Morning Session

Wednesday, October 30, 10:00-10:40 Professor Andzej Schinzel (PAN) On integer-valued polynomials

Wednesday, October 30, 11:00-11:40 Professor Katsuya Miyake (emeritus professor of Tokyo Metropolitan Univ.) Complex multiplication in the sense of Abel

Wednesday, October 30, 13:30- Excursion to Karatsu
Wednesday, October 30, 18:30-21:00 Reception at Fukuoka Garden Palace

Thursday, October 31, 2013

Morning Session

Thursday, October 31, 10:00-10:40 Professor Kohji Matsumoto (Nagoya Univ.) A joint composite hybrid strong universality theorem

Thursday, October 31, 11:00-11:30 Professor Hiroki Aoki (Tokyo Univ. of Sci.) On Jacobi forms of fractional weight

Afternoon Session

Thursday, October 31, 13:30-14:10 Professor Shigeki Egami (Shibaura Inst. of Tech.) On analytic continuation of some Dirichlet series

Thursday, October 31, 14:30-15:10 Professor X. M. Ren (Shandong Univ.) Asymptotic expansions of Voronoi's summation formulas for $\mathrm{GL}(m)$ and applications

Thursday, October 31, 15:30-15:45 Ms. Ade Irma Suriajaya (Nagoya Univ.) On the zeros of the second derivative of the Riemann zeta function under the Riemann hypothesis

Friday, November 1, 2013

Morning Session

Friday, November 1, 10:30-11:10 Professor Z.-W. Sun (Nanjing Univ.) Some new problems involving primes and permutations

Friday, November 1, 11:10-11:40 Closing

List of participants

Shigeki Akiyama, Hiroki Aoki, Tianxin Cai, Shigeki Egami, Shuichi Hayashida, Masanobu Kaneko, Shigeru Kanemitsu, Koichi Kawada, Yoshiyuki Kitaoka, Tomoya Kiyuna, Tatsuya Komatsu, Jianya Liu, Guangshi Lv, Kohji Matsumoto, Katsuya Miyake, Ryotaro Nagauchi, Yoshinobu Nakai, Tomohide Nakaya, Ryotaro Okazaki, Tomokazu Onozuka, Ryotaro Osanai, Xiumin Ren, Andrzej Schinzel, Zhiwei Sun, Ade Irma Suriajaya, Koji Tasaka, Isao Wakabayashi, Takahiro Wakasa, Takao Watanabe, Lin Weng, Laohui Xue, Jiong Yang, Wenguang Zhai

CONTENTS

Preface v

Program xv

On modular relations 1
Tomihiro Arai, Kalyan Chakraborty and Shigeru Kanemitsu

Figurate primes and Hilbert's 8th problem 65
Tianxin Cai, Yong Zhang and Zhongyan Shen

Statistical distribution of roots of a polynomial modulo prime powers 75
Yoshiyuki Kitaoka

A survey on the theory of universality for zeta and L-functions 95
Kohji Matsumoto

Complex multiplication in the sense of Abel 145
Katsuya Miyake

Problems on combinatorial properties of primes 169
Zhi-Wei Sun

Index 189

ON MODULAR RELATIONS

TOMIHIRO ARAI
Graduate School of Advanced Technology, Kinki University,
Iizuka, Fukuoka 820-8555, Japan
E-mail: tarai@fuk.kindai.ac.jp

KALYAN CHAKRABORTY
Harish-Chandra Research Institute,
Chhatnag Road, Jhunsi, Allahabad 211019, India
E-mail: kalyan@hri.res.in

SHIGERU KANEMITSU
Graduate School of Advanced Technology, Kinki University,
Iizuka, Fukuoka 820-8555, Japan
E-mail: kanemitu@fuk.kindai.ac.jp

1. Introduction

In this paper, we use a special case of the main formula in [MR] which we will further specify in the subsequent sections to give a large class of intriguing identities, modular relations, thus supplementing the coming book [MR]. Save for the last a few sections, the results are new and published for the first time.

This paper constitutes part of the first author's PhD thesis to be submitted to Kinki University.

In the first half of the paper, we confine ourselves to the unprocessed modular relations for Dedekind zeta-functions. Here we mean by the unprocessed modular relations which are not processed, i.e. there is no processing gamma factor involved and just the functional equation. This includes the treatment of transformation formulas for Lambert series as a modular relation [MR, Chapter 4].

We shall also resurrect the long-forgotten work of Koshlyakov on Dedekind zeta-functions for the rational and quadratic fields [KoshI], [KoshII], [KoshIII]. Since the Dedekind zeta-functions satisfy the functional equation with multiple gamma factors, our theorems conveniently cover many special cases of arithmetical functions including the divisor functions as well as the coefficients of zeta-functions associated to binary quadratic forms. The (positive) definite binary quadratic forms correspond to imaginary quadratic fields and indefinite ones correspond to real quadratic fields. Cf. [AAW1], [AAW2], [Coh], etc. in this regard.

Convention Thus whenever we speak of the Dedekind zeta-function of an imaginary quadratic field, we are also referring to the class of zeta-functions satisfying the Hecke type functional equation, i.e. the zeta-function associated to a positive definite quadratic form (Epstein zeta-function) or the zeta-function associated to a cusp form (or the degenerate case of the product of two zeta-functions). In the case of real quadratic fields, we also refer to the zeta-functions associated to indefinite quadratic forms or the generating function of the divisor problem.

Hitherto, there have been many modular relations studied most extensively in the past beyond the unprocessed case. They include the Fourier-Bessel expansion [MR, Chapter 4] and the Ewald expansion [MR, Chapter 5], which are highlighted.

Since resurrection of Koshlyakov' work includes the elucidation of integrated modular relations or modular relations in integral form, we are analytically continued to the Riesz sums, being integrals, which has been expounded in [MR, Chapter 6] and [WW].

To describe the modular relation we prepare the notation.

Let $\{\lambda_k\}_{k=1}^{\infty}, \{\mu_k\}_{k=1}^{\infty}$ be increasing sequences of positive numbers tending to ∞, and let $\{\alpha_k\}_{k=1}^{\infty}, \{\beta_k\}_{k=1}^{\infty}$ be complex sequences. We form the Dirichlet series

$$\varphi(s) = \sum_{k=1}^{\infty} \frac{\alpha_k}{\lambda_k^s}, \tag{1.1}$$

$$\psi(s) = \sum_{k=1}^{\infty} \frac{\beta_k}{\mu_k^s} \tag{1.2}$$

and suppose that they have finite abscissas of absolute convergence $\sigma_\varphi, \sigma_\psi$, respectively.

For the set of coefficients

$$\Delta = \begin{pmatrix} \{(a_j, A_j)\}_{j=1}^{n}; \{(1-a_j, A_j)\}_{j=n+1}^{p} \\ \{(b_j, B_j)\}_{j=1}^{m}; \{(1-b_j, B_j)\}_{j=m+1}^{q} \end{pmatrix} \in \Omega, \tag{1.3}$$

where $\in \Omega$ means that the parameters in capital letters are positive numbers and those in small letters are complex numbers, we introduce the (processing) gamma factor

$$\Gamma(w \mid \Delta) = \frac{\prod_{j=1}^{m} \Gamma(b_j + B_j w) \prod_{j=1}^{n} \Gamma(a_j - A_j w)}{\prod_{j=n+1}^{p} \Gamma(a_j + A_j w) \prod_{j=m+1}^{q} \Gamma(b_j - B_j w)} \quad (A_j, B_j > 0). \tag{1.4}$$

The (Fox) H-function is defined by ($0 \le n \le p$, $0 \le m \le q$, $A_j, B_j > 0$):

$$\begin{aligned} &H_{p,q}^{m,n}\left(z \middle| \Delta\right) \\ &= H_{p,q}^{m,n}\left(z \,\middle|\, \begin{matrix} (1-a_1, A_1), \dots, (1-a_n, A_n), (a_{n+1}, A_{n+1}), \dots, (a_p, A_p) \\ (b_1, B_1), \dots, (b_m, B_m), (1-b_{m+1}, B_{m+1}), \dots, (1-b_q, B_q) \end{matrix}\right) \\ &= \frac{1}{2\pi i} \int_L \frac{\prod_{j=1}^{m} \Gamma(b_j + B_j s) \prod_{j=1}^{n} \Gamma(a_j - A_j s)}{\prod_{j=n+1}^{p} \Gamma(a_j + A_j s) \prod_{m+1}^{q} \Gamma(b_j - B_j s)} z^{-s} \mathrm{d}s, \end{aligned} \tag{1.5}$$

where the path L is subject to the poles separation conditions similar to the ones given below.

We include the definition of the Meijer G-function in the $H \to G$ formula, which is often used without notice:

$$\begin{aligned} &H_{p,q}^{m,n}\left(z \,\middle|\, \begin{matrix} (a_1, \frac{1}{C}), \dots, (a_n, \frac{1}{C}), (a_{n+1}, \frac{1}{C}), \dots, (a_p, \frac{1}{C}) \\ (b_1, \frac{1}{C}), \dots, (b_m, \frac{1}{C}), (b_{m+1}, \frac{1}{C}), \dots, (b_q, \frac{1}{C}) \end{matrix}\right) \\ &= C\, G_{p,q}^{m,n}\left(z^C \,\middle|\, \begin{matrix} a_1, \dots, a_n, a_{n+1}, \dots, a_p \\ b_1, \dots, b_m, b_{m+1}, \dots, b_q \end{matrix}\right) \quad (C > 0). \end{aligned} \tag{1.6}$$

The case $C = 1$ is the definition of the Meijer G-function.

We suppose the existence of the meromorphic function χ satisfying the functional equation with r a real number,

$$\chi(s) = \begin{cases} \prod_{j=1}^{M} \Gamma(d_j + D_j s)\, \varphi(s), & \mathrm{Re}\,(s) > \sigma_\varphi \\ \prod_{j=1}^{\tilde{N}} \Gamma(e_j + E_j(r-s))\, \psi(r-s), & \mathrm{Re}\,(s) < r - \sigma_\psi \end{cases} \tag{1.7}$$

and having a finite number of poles s_k ($1 \le k \le L$), where

$$D_j, E_j > 0.$$

In the w-plane we take two deformed Bromwich paths

$$L_1(s) : \gamma_1 - i\infty \to \gamma_1 + i\infty, \quad L_2(s) : \gamma_2 - i\infty \to \gamma_2 + i\infty \quad (\gamma_2 < \gamma_1)$$

such that they squeeze a compact set $\mathcal{S}$ with boundary $\mathcal{C}$ for which $s_k \in \mathcal{S}$ $(1 \le k \le L)$.

Under these conditions the χ-function, *key-function*, $\mathrm{X}(z, s \,|\, \Delta)$ is usually defined in our theory as

$$\mathrm{X}(z, s \,|\, \Delta) = \frac{1}{2\pi i} \int_{L_1(s)} \Gamma(w - s \,|\, \Delta)\chi(w)\, z^{-(w-s)} dw \tag{1.8}$$

where $\Gamma(s \,|\, \Delta)$ is the processing gamma factor (1.4). In the first half of this paper we consider the case where there is no processing gamma factor, $\Gamma(s \,|\, \Delta) = 1$ and we write $\mathrm{X}(z, s) = \mathrm{X}(z, s \,|\, \emptyset)$.

1.1. *The* $H_{0,M}^{M,0} \leftrightarrow H_{0,\tilde{N}}^{\tilde{N},0}$ *formula*

We borrow the following unprocessed modular relation (first stated in [Ts]), with the H-function in (1.5).

Theorem 1.1.

$$z^s \mathrm{X}(z, s) \tag{1.9}$$

$$= \begin{cases} \displaystyle\sum_{k=1}^{\infty} \frac{\alpha_k}{\lambda_k{}^s} H_{0,\,M}^{M,0}\left(z\lambda_k \,\middle|\, \begin{matrix} - \\ \{(d_j + D_j s, D_j)\}_{j=1}^M \end{matrix} \right) \\ \qquad\qquad \textit{if } L_1(s) \textit{ can be taken to the right of } \sigma_\varphi, \\ \displaystyle\sum_{k=1}^{\infty} \frac{\beta_k}{\mu_k{}^{r-s}} H_{0,\tilde{N}}^{\tilde{N},0}\left(\frac{\mu_k}{z} \,\middle|\, \begin{matrix} - \\ \{(e_j + E_j(r-s), E_j)\}_{j=1}^{\tilde{N}} \end{matrix} \right) \\ \qquad + \displaystyle\sum_{k=1}^{L} \mathrm{Res}\Big(\chi(w)\, z^{s-w}, w = s_k\Big) \\ \qquad\qquad \textit{if } L_2(s) \textit{ can be taken to the left of } r - \sigma_\psi \end{cases}$$

is equivalent to the functional equation (1.7).

Convention In most of the formulas, the condition $z > 0$ can be extended eventually to $\mathrm{Re}\, z > 0$. In what follows we understand the proviso $z > 0$ in this wider sense whenever possible.

2. Dedekind zeta-function

We consider the Dedekind zeta-function of a number field Ω with degree $[\Omega : \mathbb{Q}] = \varkappa = r_1 + 2r_2$ (in standard notation; r_1 and $2r_2$ indicates the number of real and imaginary conjugates of Ω, respectively) and discriminant Δ. Let

$$A = \frac{2^{r_2}\pi^{\varkappa/2}}{\sqrt{|\Delta|}} \tag{2.1}$$

and let $r = r_1 + r_2 - 1$ denote the rank of the unit group. The Dedekind zeta-function $\zeta_\Omega(s)$ of the number field Ω is defined by

$$\zeta_\Omega(s) = \sum_{0\neq\mathfrak{a}\subset\Omega} \frac{1}{(\mathrm{N}\mathfrak{a})^s} = \sum_{k=1}^{\infty} \frac{F(k)}{k^s} \tag{2.2}$$

for $\sigma = \mathrm{Re}\, s > 1$, where $\mathfrak{a}$ runs through all non-zero ideals of Ω and $F(k) = F_\Omega(k)$ indicates the number of ideals of norm k, sometimes called "Idealfunction". Hence, with $\alpha_k = \beta_k = F(k)$ and $\lambda_k = \mu_k = Ak$, $r = 1$. $\zeta_\Omega(s)$ satisfies the *functional equation*

$$A^{-s}\,\Gamma^{r_1}\Big(\frac{s}{2}\Big)\,\Gamma^{r_2}(s)\,\zeta_\Omega(s) = A^{-(1-s)}\,\Gamma^{r_1}\Big(\frac{1-s}{2}\Big)\,\Gamma^{r_2}(1-s)\,\zeta_\Omega(1-s), \tag{2.3}$$

and is continued to a meromorphic function with a pole at $s = 1$ with residue (cf. (5.12) for $\varkappa \le 2$)

$$\lambda h = \frac{2^{r+1}\pi^{r_2}Rh}{w\sqrt{|\Delta|}}. \tag{2.4}$$

Here w, h and R are the number of roots of unity in Ω, the class number and the regulator respectively.

We define the meromorphic function in (1.7) as

$$\chi(s) = \Gamma^{r_1}\Big(\frac{s}{2}\Big)\,\Gamma^{r_2}(s)\varphi(s), \tag{2.5}$$

where

$$\varphi(s) = A^{-s}\zeta_\Omega(s) = \sum_{k=1}^{\infty} \frac{\alpha(k)}{(Ak)^s}. \tag{2.6}$$

The functional equation (1.7) in this case reads

$$\begin{aligned}\chi(s) &= \Gamma^{r_1}\Big(\frac{1}{2}s\Big)\Gamma^{r_2}(s)\varphi(s) = \Gamma^{r_1}\Big(\frac{1}{2} - \frac{1}{2}s\Big)\Gamma^{r_2}(1-s)\varphi(1-s) \\ &= \chi(1-s)\end{aligned} \tag{2.7}$$

Table 1. Invariants of a number field.

symbol	meaning
$\varkappa = r_1 + 2r_2$	degree of Ω
r_1	number of real conjugates of Ω
$2r_2$	number of imaginary conjugates of Ω
$r = r_1 + r_2 - 1$	rank of the unit group of Ω
Δ	discriminant of Ω
h	class number
R	regulator
w	number of roots of unity in Ω
$\zeta_\Omega(s)$	Dedekind zeta-function of Ω
$F(k) = F_\Omega(k)$	number of ideals whose norm is k
$A = \frac{2^{r_2}\pi^{\varkappa/2}}{\sqrt{\lvert\Delta\rvert}}$	coefficient of functional equation for $\zeta_\Omega(s)$
$\lambda h = \frac{2^{r+1}\pi^{r_2}Rh}{w\sqrt{\lvert\Delta\rvert}}$	residue at $s = 1$ of $\zeta_\Omega(s)$

and $\chi(s)$ has poles at $s = 1, 0$ (and possibly at negative integer points, which will not be relevant to us).

(2.7) is a special case of (1.7) with $d_j = 0, e_j = 0$, $D_j = 1\,(1 \le j \le r_1)$, $D_j = \frac{1}{2}\,(r_1 + 1 \le j \le r_1 + r_2 = M)$ and $D_j = 1\,(1 \le j \le r_1)$, $E_j = \frac{1}{2}\,(r_1 + 1 \le j \le r_1 + r_2 = \tilde{N} = M)$ and $r = 1$.

Hence Theorem 1.1 reads

Theorem 2.1.

$$z^s \mathrm{X}(z,s) \tag{2.8}$$

$$= \begin{cases} \displaystyle\sum_{k=1}^{\infty} \frac{\alpha_k}{(Ak)^s} H_{0,\,M}^{M,0}\left(z\lambda_k \,\middle|\, \begin{matrix} - \\ \{(\frac{s}{2},\frac{1}{2})\}_{j=1}^{r_1}, \{(s,1)\}_{j=r_1+1}^{r_2} \end{matrix} \right) \\ \qquad\qquad \textit{if } L_1(s) \textit{ can be taken to the right of } \sigma_\varphi, \\ \\ \displaystyle\sum_{k=1}^{\infty} \frac{\alpha_k}{(Ak)^{1-s}} H_{0,\tilde{N}}^{\tilde{N},0}\left(\frac{\mu_k}{z} \,\middle|\, \begin{matrix} - \\ \{(\frac{1-s}{2},\frac{1}{2})\}_{j=1}^{r_1}, \{(1-s,1)\}_{j=r_1+1}^{r_2} \end{matrix} \right) \\ \qquad + \displaystyle\sum_{k=1}^{L} \mathrm{Res}\Big(\chi(w)\, z^{s-w}, w = s_k \Big) \\ \qquad\qquad \textit{if } L_2(s) \textit{ can be taken to the left of } 1 - \sigma_\psi. \end{cases}$$

Using the notation of Koshylakov, we introduce the X-function by

$$X_{r_1,r_2}(x) = \frac{1}{2\pi i}\int_{(c)} \Gamma^{r_1}\left(\frac{s}{2}\right)\Gamma^{r_2}(s)x^{-s}\,\mathrm{d}s, \quad c>0, \quad \sigma>0, \tag{2.9}$$

for $x > 0$.

Or in terms of the H-function in (1.5), it is

$$X(x) = X_{r_1,r_2}(x) = H_{0,\,\varkappa}^{\varkappa,0}\left(x \,\middle|\, \begin{matrix} - \\ \{(0,\frac{1}{2})\}_{j=1}^{r_1}, \{(0,1)\}_{j=r_1+1}^{r_2} \end{matrix}\right). \tag{2.10}$$

Corollary 2.1. *Theorem 2.1 asserts that in the case* $\varkappa \le 2$

$$\sum_{k=1}^{\infty} F_\Omega(k) X_{r_1,r_2}(Ak\rho) = \mathrm{P}(x) + \sum_{k=1}^{\infty} F_\Omega(k) X_{r_1,r_2}(Ak\rho^{-1}), \tag{2.11}$$

where $\mathrm{P}(x) = R_0 + R_1$ *is the residual function and*

$$R_j = \mathrm{Res}\left(\chi(s)z^{-s}, s=j\right), \quad j=0,1. \tag{2.12}$$

3. Glossary of formulas

We assemble here those formulas for H- and G-functions that will be often used subsequently and in particular those which amount to Bessel functions. For details we refer to [MR, Chapter 2].

The J-Bessel function $J_s(z)$ has the integral representation [ErdH, III,(34) p.21]

$$J_\nu(x) = \frac{1}{2}\frac{1}{2\pi i}\int_{(c)} \frac{\Gamma(\frac{1}{2}\nu + \frac{1}{2}z)}{\Gamma(1+\frac{1}{2}\nu - \frac{1}{2}z)}\left(\frac{x}{2}\right)^{-z}\mathrm{d}z \tag{3.1}$$

$$= \frac{1}{2}H_{0,2}^{1,0}\left(\frac{x}{2} \,\middle|\, \begin{matrix} - \\ \left(\frac{\nu}{2},\frac{1}{2}\right), \left(-\frac{\nu}{2},\frac{1}{2}\right) \end{matrix}\right)$$

for $-\mathrm{Re}\,\nu < c < 1$ ($x > 0$, which may read $\mathrm{Re}\,x > 0$). We note that on the left-side of [ErdH, III,(34) p.21] should read $4\pi i J_\nu(x)$ and that without the restriction $-\mathrm{Re}\,\nu < c < 1$, (3.1) still makes sense, but it need not represent the Bessel function.

We note that in general the Bessel functions, say F_s reduce in the case of s being a half-integer $\nu + \frac{1}{2}$, to an elementary function ([ErdT, II]); cf. (3.2), (3.7). $J_s(z)$ reduces to

$$J_{\nu+1/2}(z) = (-1)^\nu \frac{1}{\sqrt{\frac{\pi}{2}z}} z^{\nu+1}\left(\frac{\mathrm{d}}{z\mathrm{d}z}\right)^\nu \frac{\sin z}{z} \tag{3.2}$$

$$= \frac{1}{\sqrt{\frac{\pi}{2}z}}\left(\sin\left(z-\frac{\pi}{2}n\right)\sum_{m=0}^{[\frac{n}{2}]}(-1)^m\binom{n+1/2}{2m}(2z)^{-2m}\right.$$
$$\left.+\cos\left(z-\frac{\pi}{2}n\right)\sum_{m=0}^{[\frac{n-1}{2}]}(-1)^m\binom{n+1/2}{2m+1}(2z)^{-2m-1}\right),$$

where $[x]$ indicates the integer part of x, i.e. the greatest integer not exceeding x.

Associated to the J-Bessel function is its complex version ($\nu \ni \mathbb{Z}$)

$$I_\nu(z) = e^{-\frac{\pi}{2}i\nu}J_\nu(ze^{-\frac{\pi}{2}i\nu}) \tag{3.3}$$

called the modified Bessel function of the first kind. It is a solution to

$$x^2\frac{\mathrm{d}^2w}{\mathrm{d}x^2} + x\frac{\mathrm{d}w}{\mathrm{d}x} - (x^2+\nu^2)w = 0, \tag{3.4}$$

$$-\pi Y_s(x) = \frac{1}{2\pi i}\int_{(c)}\Gamma(z)\Gamma(z-s)\cos\pi(z-s)\left(\frac{x}{2}\right)^{s-2z}\mathrm{d}z, \tag{3.5}$$

for $0 \le \mathrm{Re}(s) < c < \mathrm{Re}(s) + \frac{3}{4} < \frac{3}{2}$ [ErdH].

The K-Bessel function $K_s(z)$ may be defined by ([Vista II, (6.14), p.107])

$$K_\nu(z) = \frac{1}{2}\left(\frac{z}{2}\right)^\nu\int_0^\infty e^{-t-\frac{z^2}{4t}}t^{-\nu-1}\mathrm{d}t, \quad \mathrm{Re}(\nu) > -\frac{1}{2}, \quad |\arg z| < \frac{\pi}{4}. \tag{3.6}$$

$K_s(z)$ reduces ([ErdH, (40), p.10]) to

$$K_{\nu+\frac{1}{2}}(z) = \left(\frac{\pi}{2z}\right)^{1/2}e^{-z}\sum_{m=0}^{\nu}(2z)^{-m}\frac{\Gamma(\nu+m+1)}{m!\,\Gamma(\nu+1+m)}. \tag{3.7}$$

A special case of (3.7) which is often used without notice is

$$K_{\pm\frac{1}{2}}(z) = \sqrt{\frac{\pi}{2z}}e^{-z}. \tag{3.8}$$

The K-Bessel function is related to other Bessel functions by

$$Y_\kappa(iz) = e^{\frac{\pi i(\kappa+1)}{2}}I_\kappa(z) - \frac{2}{\pi}e^{-\frac{\pi i\kappa}{2}}K_\kappa(z), \qquad -\pi < \arg z \le \frac{\pi}{2}, \tag{3.9}$$

and for $\kappa = 1$ we see that

$$Y_1(iz) = e^{\pi i}I_1(z) - \frac{2}{\pi}e^{-\frac{\pi i}{2}}K_1(z) = -e^{-\frac{\pi i}{2}}J_1(iz) - \frac{2}{\pi}e^{-\frac{\pi i}{2}}K_1(z),$$
$$K_1(z) = \frac{\pi}{2}\left(-J_1(iz) - e^{\frac{\pi i}{2}}Y_1(iz)\right)$$

i.e.

$$K_1(-iz) = \frac{\pi}{2}\left(-J_1(z) - e^{\frac{\pi i}{2}}Y_1(z)\right), \quad K_1(iz) = \frac{\pi}{2}\left(-J_1(-z) - e^{\frac{\pi i}{2}}Y_1(-z)\right). \tag{3.10}$$

In addition to the $H \to G$ formula (1.6) above, we often use other reduction formulas whose expounding may be found in [MR, Chapter 2].

The *reduction-augmentation formula*

$$\begin{aligned} &H\left(z \,\middle|\, \Delta \oplus \begin{pmatrix} (c,C); & - \\ - & ;(c,C) \end{pmatrix}\right) \\ &= H\left(z \,\middle|\, \Delta \oplus \begin{pmatrix} - & ;(c,C) \\ (c,C); & - \end{pmatrix}\right) \\ &= H(z|\, \Delta). \end{aligned} \tag{3.11}$$

Theorem 3.1 (Reciprocity formula). *The reciprocity formula for the gamma function and Euler's identity lead to*

$$\begin{aligned} &H\left(z \,\middle|\, \Delta \oplus \begin{pmatrix} -;(c,C) \\ -;(c,C) \end{pmatrix}\right) \\ &= \frac{1}{2\pi i}\left\{e^{c\pi i}\, H\left(e^{-C\pi i}z|\, \Delta\right) - e^{-c\pi i}\, H\left(e^{C\pi i}z|\, \Delta\right)\right\}. \end{aligned} \tag{3.12}$$

In the traditional notation, (3.12) *reads*

$$\begin{aligned} &H^{m,n}_{p+1,q+1}\left(z \,\middle|\, \begin{matrix} \{(a_j,A_j)\}_{j=1}^n, \{(a_j,A_j)\}_{j=n+1}^p, (c,C) \\ \{(b_j,B_j)\}_{j=1}^m, \{(b_j,B_j)\}_{j=m+1}^q, (c,C) \end{matrix}\right) \\ &= \frac{1}{2\pi i}\left\{e^{c\pi i}\, H^{m,n}_{p,q}\left(e^{-C\pi i}z \,\middle|\, \begin{matrix} \{(a_j,A_j)\}_{j=1}^n, \{(a_j,A_j)\}_{j=n+1}^p \\ \{(b_j,B_j)\}_{j=1}^m, \{(b_j,B_j)\}_{j=m+1}^q \end{matrix}\right)\right. \\ &\quad \left. - e^{-c\pi i}\, H^{m,n}_{p,q}\left(e^{C\pi i}z \,\middle|\, \begin{matrix} \{(a_j,A_j)\}_{j=1}^n, \{(a_j,A_j)\}_{j=n+1}^p \\ \{(b_j,B_j)\}_{j=1}^m, \{(b_j,B_j)\}_{j=m+1}^q \end{matrix}\right)\right\}, \end{aligned} \tag{3.13}$$

which entails the following formula for G-functions:

$$\begin{aligned} &G^{m,n}_{p+1,q+1}\left(z \,\middle|\, \begin{matrix} a_1,\ldots,a_n,a_{n+1},\ldots,a_p,c \\ b_1,\ldots,b_m,b_{m+1},\ldots,b_q,c \end{matrix}\right) \\ &= \frac{1}{2\pi i}\left\{e^{c\pi i}\, G^{m,n}_{p,q}\left(e^{-\pi i}z \,\middle|\, \begin{matrix} a_1,\ldots,a_n,a_{n+1},\ldots,a_p \\ b_1,\ldots,b_m,b_{m+1},\ldots,b_q \end{matrix}\right)\right. \\ &\quad \left. - e^{-c\pi i}\, G^{m,n}_{p,q}\left(e^{\pi i}z \,\middle|\, \begin{matrix} a_1,\ldots,a_n,a_{n+1},\ldots,a_p \\ b_1,\ldots,b_m,b_{m+1},\ldots,b_q \end{matrix}\right)\right\}. \end{aligned} \tag{3.14}$$

Corollary 3.1. *We have*

$$\begin{aligned} &H^{m,n}_{p+1,q+1}\left(z \,\middle|\, \begin{matrix} \{(a_j,A_j)\}_{j=1}^{n}, \{(a_j,A_j)\}_{j=n+1}^{p}, (c+1,C) \\ \{(b_j,B_j)\}_{j=1}^{m}, \{(b_j,B_j)\}_{j=m+1}^{q}, (c+1,C) \end{matrix}\right) \\ &= -H^{m,n}_{p+1,q+1}\left(z \,\middle|\, \begin{matrix} \{(a_j,A_j)\}_{j=1}^{n}, \{(a_j,A_j)\}_{j=n+1}^{p}, (c,C) \\ \{(b_j,B_j)\}_{j=1}^{m}, \{(b_j,B_j)\}_{j=m+1}^{q}, (c,C) \end{matrix}\right), \end{aligned} \tag{3.15}$$

which includes the following equation for G-functions:

$$\begin{aligned} &G^{m,n}_{p+1,q+1}\left(z \,\middle|\, \begin{matrix} a_1,\ldots,a_n,a_{n+1},\ldots,a_p,c+1 \\ b_1,\ldots,b_m,b_{m+1},\ldots,b_q,c+1 \end{matrix}\right) \\ &= -G^{m,n}_{p+1,q+1}\left(z \,\middle|\, \begin{matrix} a_1,\ldots,a_n,a_{n+1},\ldots,a_p,c \\ b_1,\ldots,b_m,b_{m+1},\ldots,b_q,c \end{matrix}\right). \end{aligned} \tag{3.16}$$

The (Hecke) *gamma transform* [Vista II, p.25] reads for $\lambda > 0$ and $\sigma > 0$

$$\Gamma(s)\lambda^{-s} = \int_0^\infty t^s e^{-\lambda t}\,\frac{dt}{t} \tag{3.17}$$

which is a special case of

$$G^{1,0}_{0,1}\left(z \,\middle|\, \begin{matrix} - \\ a \end{matrix}\right) = z^a\, e^{-z}. \tag{3.18}$$

Also we have

$$G^{1,0}_{0,2}\left(z \,\middle|\, \begin{matrix} - \\ a,b \end{matrix}\right) = z^{\frac{1}{2}(a+b)} J_{a-b}(2\sqrt{z}) \tag{3.19}$$

whose special cases read

$$\begin{aligned} G^{1,0}_{0,2}\left(z \,\middle|\, \begin{matrix} - \\ a, a+\frac{1}{2} \end{matrix}\right) &= \frac{z^a}{\sqrt{\pi}}\cos(2\sqrt{z}), \\ G^{1,0}_{0,2}\left(z \,\middle|\, \begin{matrix} - \\ a+\frac{1}{2}, a \end{matrix}\right) &= \frac{z^a}{\sqrt{\pi}}\sin(2\sqrt{z}). \end{aligned} \tag{3.20}$$

There is a hierarchy of formulas. The seemingly highest is

$$G^{1,1}_{1,2}\left(z \,\middle|\, \begin{matrix} a \\ b,c \end{matrix}\right) = z^b\,\frac{\Gamma(1-a+b)}{\Gamma(1-c+b)}\,{}_1F_1\left(\begin{matrix} 1-a+b \\ 1-c+b \end{matrix}; -z\right). \tag{3.21}$$

It reduces, in view of (3.33), to

$$G^{1,1}_{1,2}\left(z \,\middle|\, \begin{matrix} 1 \\ a,0 \end{matrix}\right) = \Gamma(a) - \Gamma(a,z). \tag{3.22}$$

We note that neither (3.21) or its complement (3.30) appears in the table of formulas for G-functions in [ErdH, pp.216-222] but they do appear as integral representations (Mellin-Barnes integrals, or the G-functions), [ErdH, (4), p.256] and [ErdH, (6), p.256], respectively.

A special case of (3.21) is stated as [ErdH, (4), p.256] (or [PBM, 8.4.45.1, p.715])

$$G_{1,2}^{1,1}\left(z\,\middle|\,\begin{matrix}1-a\\0,1-b\end{matrix}\right)=\frac{\Gamma(b)}{\Gamma(a)}\,{}_1F_1\left(\begin{matrix}a\\b\end{matrix};-z\right)=\frac{\Gamma(b)}{\Gamma(a)}\,\Phi(a,b;-z). \tag{3.23}$$

We show that (3.21) is indeed equivalent to (3.23). For writing $\alpha=1-a+b$, $\beta=1-c+b$, we may write the left-hand side as $G_{1,2}^{1,1}\left(z\,\middle|\,\begin{matrix}1-\alpha+b\\b,1-\beta+b\end{matrix}\right)$. Hence by the translation, it becomes $G_{1,2}^{1,1}\left(z\,\middle|\,\begin{matrix}1-\alpha\\0,1-\beta\end{matrix}\right)$, whence (3.23) implies (3.21). The reverse implication being clear, this completes the proof.

The *inverse Heaviside integral* formula ([Vista I, p,108], [Vista II, p.98]) reads

$$G_{0,2}^{2,0}\left(z\,\middle|\,\begin{matrix}-\\a,b\end{matrix}\right)=2\,z^{\frac{1}{2}(a+b)}K_{a-b}\big(2\sqrt{z}\,\big), \tag{3.24}$$

where $K_\kappa(z)$ be the Bessel function of the second kind defined by (3.6).

(3.24) is a companion to

$$G_{1,1}^{1,1}\left(z\,\middle|\,\begin{matrix}a\\b\end{matrix}\right)=\Gamma(1-a+b)\,z^b\,(z+1)^{a-b-1} \tag{3.25}$$

and the beta transform (or the beta integral) [MR, p.134] is a special case thereof, which leads to the Fourier-Bessel expansion [MR, Chapter 4].

[ErdH, (6), p.216] reads

$$G_{1,2}^{2,0}\left(z\,\middle|\,\begin{matrix}a\\b,c\end{matrix}\right)=z^{\frac{1}{2}(b+c-1)}e^{-\frac{1}{2}z}\,W_{\kappa,\mu}(z), \tag{3.26}$$

where $W_{\kappa,\mu}$ is the *Whittaker function* with indices $\kappa=\frac{1}{2}(b+c+1)-a$, $\mu=\frac{1}{2}(b-c)$ and [ErdH, (8), p.216] reads

$$G_{1,2}^{2,1}\left(z\,\middle|\,\begin{matrix}a\\b,c\end{matrix}\right)=\Gamma(1-a+b)\,\Gamma(1-a+c)\,z^{\frac{1}{2}(b+c-1)}e^{\frac{1}{2}z}\,W_{\kappa,\mu}(z), \tag{3.27}$$

where the indices of the Whittaker function are $\kappa=a-\frac{1}{2}(b+c+1)$, $\mu=\frac{1}{2}(b-c)$.

In view of [ErdH, (2), p.264]

$$W_{\kappa,\mu}(z)=e^{-\frac{1}{2}z}z^{\frac{1}{2}c}U(a,c;z),\quad a=\frac{1}{2}-\kappa+\mu,\ c=2\mu+1, \tag{3.28}$$

(3.26) is equi-vocal to

$$G_{1,2}^{2,0}\left(z \,\middle|\, \begin{matrix} a \\ b, c \end{matrix}\right) = e^{-z} z^b U\big(a-c, b-c+1, z\big), \tag{3.29}$$

where $U(a,c;z)$ also denoted by $\Psi(a,c;z)$ is the confluent hypergeometric function of the second kind which often appears in §9:

$$U(a,c;z) = \Psi(a,c;z).$$

On the other hand, (3.27) is equivalent to

$$\begin{aligned} &G_{1,2}^{2,1}\left(z \,\middle|\, \begin{matrix} a \\ b, c \end{matrix}\right) \\ &= \Gamma(1-a+b)\,\Gamma(1-a+c)\, z^b U\big(b-a+1, b-c+1, z\big). \end{aligned} \tag{3.30}$$

In view of

$$\Gamma(a,z) = e^{-z} U(1-a, 1-a; z), \tag{3.31}$$

which is [ErdH, (21), p.266], (3.29) with $a=1, c=0$ and a for b reduces to

$$G_{1,2}^{2,0}\left(z \,\middle|\, \begin{matrix} 1 \\ a, 0 \end{matrix}\right) = \Gamma\big(a,z\big). \tag{3.32}$$

We also note

$$\gamma(a,z) = \Gamma(a) - \Gamma(a,z) = a^{-1} z^a \Phi(a, a+1; z), \tag{3.33}$$

which is [ErdH, (22), p.266] and

$$\Gamma(a,z) = e^{-z} z^a U(1, a+1; z), \tag{3.34}$$

which is [PBM, 7.11.4.11, p.584].

Finally we mention generalized Bessel and Voronoĭ functions. Both [BeI] and [BeII] treat the functional equation

$$\Gamma^m(s)\varphi(s) = \Gamma^m(r-s)\psi(r-s) \tag{3.35}$$

for m a positive integer, extending Hecke's functional equation, cf. (5.34). In [BeI] the equivalence is proved of (3.35) to the modular relation ([BeI, Theorem 1]),

$$\sum_{k=1}^{\infty} \alpha_k E_m(\lambda_k x) = x^{-r} \sum_{k=1}^{\infty} \beta_k E_m(\mu_k/x) + \mathrm{P}(x), \tag{3.36}$$

where $E_m(x)$ indicates *Voronoĭ's function*

$$E_m(x) = \frac{1}{2\pi i}\int_{(c)} \Gamma^m(s)x^{-s}\,\mathrm{d}s = G^{m,0}_{0,m}\left(z\,\middle|\, \begin{matrix} - \\ \underbrace{0,\dots,0}_{m} \end{matrix}\right) \quad (x>0), \tag{3.37}$$

and to the Riesz sum of order $\varkappa$:

$$\frac{1}{\Gamma(\varkappa+1)}\sum_{\lambda_k\le x}{}' \alpha_k(x-\lambda_k)^{\varkappa}$$

involving the generalized Bessel function $K_\nu(x;\mu;m)$ ([BeI, Definition 4])

$$\begin{aligned} K_\varkappa(z;r,m) &= \frac{z^\varkappa}{2\pi i}\int_L \frac{\Gamma(s)}{\Gamma(\varkappa+1-s)}\frac{\Gamma^{m-1}(s)}{\Gamma^{m-1}(r-s)}\left(\frac{z^2}{2^{2m}}\right)^{-s}\mathrm{d}s \\ &= z^\varkappa H^{m,0}_{0,m}\left(\left(\frac{z}{2^m}\right)^2 \,\middle|\, \begin{matrix} - \\ \underbrace{(0,1),\cdots,(0,1)}_{m},(-\varkappa,1),\underbrace{(1-r,1),\cdots,(1-r,1)}_{m} \end{matrix}\right). \end{aligned} \tag{3.38}$$

Thus these functions appear in the case of powers of zeta-functions, which we shall not treat here and refer to [MR, §9.2]. We state, however, one particular case (9.44). We remark that Bellman had already introduced the same function in 1949 in [Bel2], [Bel3]. In [Bel4, (1b)], he refers to the Siegel integral, which has been introduced by Wishart [Wis] (cf. [Sa] for more details). In [Bel5] he further refers to the most general *Steen's function* $V = V(x;a_1,\cdots,a_n)$ ([St]) as:

$$\frac{1}{2\pi i}\int_0^\infty x^s V(x;a_1,\cdots,a_n)\frac{\mathrm{d}x}{x} = \Gamma(s+a_1)\cdots\Gamma(s+a_n). \tag{3.39}$$

This is nothing but the G-function formula

$$V(x;a_1,\cdots,a_n) = G^{n,0}_{0,n}\left(x\,\middle|\, \begin{matrix} - \\ a_1,\dots,a_n \end{matrix}\right). \tag{3.40}$$

4. Koshlyakov's method [KoshI]

This section will serve as fixing some notation, including the choice of the fundamental sequence (by incorporating the associated constants in them). We shall elucidate the trilogy of Koshlyalov and show that Koshlyakov's theory of Dedekind zeta-functions of quadratic fields [KoshI] reduces in the long run to Theorem 2.1.

This trilogy hitherto has not been well-known compared to with his other papers [Kosum] and [Kosvor] but is rich in contents and can be thought of as the compilation of his work. To expound it is beneficial in two respects. First, one may come to know this relatively unknown but important work, and secondly, to our benefit, we can exhibit three typical cases of the functional equation corresponding to the rational, imaginary and real quadratic field. The latter two corresponds to the zeta-functions of definite and indefinite quadratic forms, respectively. The zeta-functions of (positive) definite quadratic forms are the Epstein zeta-functions whose theory has been developed rather fully. We refer to [Te5], [Vista I, Chapter 6] and [Vista II, Chapter 6] for the definite case, and refer to [Sie3] and [Sie4] for the latter. This view point may be found in [Coh] as well as in [Dav], which refers to [LanZT] for proofs.

We follow the notation of Koshlyakov with slight deviations. The number field $\Omega = \Omega(\sqrt{\Delta})$ introduced in §2 is a rational or a quadratic field, i.e. if $\varkappa = [\Omega : \mathbb{Q}]$, then $\varkappa \le 2$ and Δ stands for the *discriminant* of Ω, so that $\Omega = \Omega(\sqrt{\Delta})$ under the convention that $\Delta = 1$ in the case of the prime field $\Omega = \mathbb{Q}$.

$$\begin{cases} \Delta = 1 & n = 1 \\ \Delta = \text{a square-free integer} & \varkappa = 2. \end{cases} \tag{4.1}$$

We write $\varkappa = r_1 + 2r_2 \le 2$ and $r = r_1 + r_2 - 1 \le 1$, the rank of the unit group of Ω.
Where

$$\varphi(s) = A^{-s}\zeta_\Omega(s) = \sum_{n=1}^{\infty} \frac{F_\Omega(n)}{(An)^s} = \begin{cases} \pi^{-\frac{s}{2}}\zeta(s) = \sum_{n=1}^{\infty} \frac{1}{(\sqrt{\pi}n)^s} & \varkappa = 1 \\ \sum_{n=1}^{\infty} \frac{F_\Omega(n)}{(\frac{2\pi}{\sqrt{|\Delta|}}n)^s} & \varkappa = 2, \Delta < 0 \\ \sum_{n=1}^{\infty} \frac{F_\Omega(n)}{(\frac{\sqrt{\pi}}{\sqrt{|\Delta|}}n)^s} & \varkappa = 2, \Delta > 0. \end{cases} \tag{4.2}$$

The functional equation in (2.7) takes the form

$$\chi(s) = \Gamma^{r_1}\Big(\frac{s}{2}\Big)\,\Gamma^{r_2}(s)\varphi(s) = \begin{cases} \Gamma\left(\frac{s}{2}\right)\pi^{-\frac{s}{2}}\zeta(s) & \varkappa = 1 \\ \Gamma(s)\sum_{n=1}^{\infty} \frac{F_\Omega(n)}{(\frac{2\pi}{\sqrt{|\Delta|}}n)^s} & \varkappa = 2, \Delta < 0 \\ \Gamma\left(\frac{s}{2}\right)^2\sum_{n=1}^{\infty} \frac{F_\Omega(n)}{(\frac{\sqrt{\pi}}{\sqrt{|\Delta|}}n)^s} & \varkappa = 2, \Delta > 0 \end{cases} \tag{4.3}$$

$$= \chi(1-s).$$

5. Koshlyakov's functions

Koshlyakov introduced several functions starting from the K-function (kernel function), $K(x) = K_{r_1,r_2}(x)$, defined by (6.18), which plays the most prominent role in his theory; along with K, also introduced are X, Y, Z, L, M, N, see Table 8.2 ([KoshI, p.122]). We shall treat some (properties) of them.

5.1. *Koshlyakov's X-functions*

For the X-function defined by (2.9), we prove the following formulas.

$$X_{1,0}(x) = \frac{1}{2\pi i}\int_{(c)} \Gamma\left(\frac{s}{2}\right) x^{-s}\, \mathrm{d}s = 2e^{-x^2} \tag{5.1}$$

$$X_{0,1}(x) = G^{1,0}_{0,1}\left(x \,\middle|\, \begin{matrix} - \\ 0 \end{matrix}\right) = e^{-x} \tag{5.2}$$

and

$$X_{2,0}(x) = \frac{1}{2\pi i}\int_{(c)} \Gamma^2\left(\frac{s}{2}\right) x^{-s}\, \mathrm{d}s = 4K_0(2x). \tag{5.3}$$

Since

$$X_{1,0}(x) = H^{1,0}_{0,1}\left(x \,\middle|\, \begin{matrix} - \\ (0, \frac{1}{2}) \end{matrix}\right),$$

we obtain by (1.6), $X_{1,0}(x) = 2G^{1,0}_{0,1}\left(x^2 \,\middle|\, \begin{matrix} - \\ 0 \end{matrix}\right)$. Then we apply (3.18) to deduce (5.1), which we apply directly to $X_{0,1}(x)$ to deduce (5.2). Finally,

$$X_{2,0}(x) = H^{2,0}_{0,2}\left(x \,\middle|\, \begin{matrix} - \\ (0, \frac{1}{2}), (0, \frac{1}{2}) \end{matrix}\right),$$

by (3.24).

We prove that the residual function $\mathrm{P}(x) = R_0 + R_1$, cf. (2.12), may be expressed as

$$R_0 + R_1 = 2^{r_1}\zeta_\Omega^{(r)}(0) - 2^{r_1}\zeta_\Omega^{(r)}(0)z^{-1}, \tag{5.4}$$

which is the *corrected form* of [KoshI, p.119, l.11 from below]. We distinguish three cases.

In the case of $\zeta(s) = \zeta_{\mathbb{Q}}(s)$, we have

$$R_0 = 2\zeta(0), \quad R_1 = -2\zeta(0)\, z^{-1}.$$

In the imaginary quadratic case we have

$$R_0 = \zeta_\Omega(0), \quad R_1 = -\zeta_\Omega(0)\, z^{-1}.$$

Finally, we turn to the real quadratic case, where the functional equation (2.7) implies that the LHS has a simple pole at $s = 1$ while the RHS has a seemingly double pole at $s = 1$, whence we must have $\zeta_\Omega(0) = 0$.

$$\pi^{-\frac{s}{2}}\Gamma^2\left(\frac{s}{2}\right)\zeta_\Omega(s)\, z^{-s} = \pi^{-\frac{s}{2}}\frac{4}{s^2}\Gamma^2\left(1+\frac{s}{2}\right)\left(\zeta'_\Omega(0)(s) + \cdots\right) z^{-s}$$

gives

$$R_0 = 4\zeta'_\Omega(0).$$

Since

$$\begin{aligned} &\pi^{-\frac{s}{2}}\Gamma^2\left(\frac{s}{2}\right)\zeta_\Omega(s)\, z^{-s} \\ &= \pi^{-\frac{1-s}{2}}\frac{4}{(s-1)^2}\Gamma^2\left(\frac{3-s}{2}\right)\left(-\zeta'_\Omega(0)(s-1) + \cdots\right) z^{-s} \\ &= \left(-\frac{4\zeta'_\Omega(0)}{s-1} + \cdots\right) z^{-s}, \end{aligned}$$

we have

$$R_1 = -4\zeta'_\Omega(0) z^{-1}.$$

Remark 5.1. In [KoshE], Koshlyakov mentions:

"On [KoshI, p. 114], the erroneous value $\zeta_\Omega(0) = -1$ of the Dedekind zeta-function for the imaginary quadratic field is given. The true value is $\zeta_\Omega(0) = -\frac{h}{2}$ with the class number h in my previous paper $\cdots$."

However, the correction is valid except for the cases of the Gauss field $\Omega = \mathbb{Q}(i)$ $(w = 4)$ and the Eisenstein field $\Omega = \mathbb{Q}(\rho)$ $(w = 6)$, where $i = e^{\frac{2\pi i}{4}}$ and $\rho = e^{\frac{2\pi i}{6}}$ indicates the piervot'nyi primitive root of unity. In our case of $\varkappa \le 2$, we may unify the values into the form

$$\zeta_\Omega(0) = -\frac{h}{w} \tag{5.5}$$

with understanding that $h = 1$, $w = 2$ in the rational case and $w = \infty$ in the real quadratic field. Of course, in the rational case, a more familiar expression for the values of the Riemann zeta-function at non-positive integral arguments is that of Euler (cf. [LWK, Proposition 5.1, p. 138]), $\zeta(0) = -\frac{B_0}{2} = -\frac{1}{2}$. Euler's formula is a consequence of the functional equation. Historically, Euler first prove the expression and conjectured the general form of the functional equation.

We are now in a position to state and prove the following theorem which corrects Koshlyakov [KoshI, (4.1)]. For that purpose we shall prove the three identities (5.7), (5.10), and (5.11) independently using the facts proved above on the concrete from of the X-functions and the residual functions. Each case has its own interest.

Theorem 5.1. *In the case of the rational or a quadratic field, the unprocessed modular relation* (2.11) *reads*

$$\sqrt{\rho}\left\{2^{r_1}\zeta_\Omega^{(r)}(0)-\sum_{k=1}^{\infty}F(k)X_{r_1,r_2}(Ak\rho)\right\}$$
$$=\frac{1}{\sqrt{\rho}}\left\{2^{r_1}\zeta_\Omega^{r}(0)-\sum_{k=1}^{\infty}F(k)X_{r_1,r_2}\left(Ak\frac{1}{\rho}\right)\right\}. \tag{5.6}$$

(2.11) leads, on incorporating (5.5), to Koshlyakov's formula [KoshI, (4.8)]:

$$\sqrt{\rho}\left\{1+2\sum_{k=1}^{\infty}e^{-\pi k^2\rho^2}\right\}=\frac{1}{\sqrt{\rho}}\left\{1+2\sum_{k=1}^{\infty}e^{-\pi k^2\frac{1}{\rho^2}}\right\}. \tag{5.7}$$

Writing ρ^2 for z, we rewrite it as

$$-\zeta(0)+\sum_{k=1}^{\infty}e^{-\pi k^2 z}=z^{-\frac{1}{2}}\left(-\zeta(0)+\sum_{k=1}^{\infty}e^{-\pi k^2\frac{1}{z}}\right)$$

or

$$\sum_{k=1}^{\infty}e^{-\pi k^2 z}+\frac{1}{2}=z^{-\frac{1}{2}}\sum_{k=1}^{\infty}\left(e^{-\pi k^2 z^{-1}}+\frac{1}{2}\right),$$

the theta-transformation formula

$$\vartheta(z)=\frac{1}{\sqrt{z}}\vartheta\left(\frac{1}{z}\right) \tag{5.8}$$

valid for $\mathrm{Re}\, z>0$, on writing

$$\vartheta(z)=\vartheta_2(z)=\sum_{n=-\infty}^{\infty}e^{-\pi n^2 z}. \tag{5.9}$$

In the imaginary quadratic case, we have

$$-\zeta_\Omega(0)+\sum_{k=1}^{\infty}F(k)e^{-\frac{2\pi}{\sqrt{|\Delta|}}z}=z^{-1}\left(-\zeta_\Omega(0)+\sum_{k=1}^{\infty}F(k)e^{-\frac{2\pi}{\sqrt{|\Delta|}}z^{-1}}\right),$$

or

$$\sqrt{\rho}\left(h+w\sum_{k=1}^{\infty}F(k)e^{-\frac{2\pi}{\sqrt{|\Delta|}}\rho}\right)=\frac{1}{\sqrt{\rho}}\left(h+w\sum_{k=1}^{\infty}F(k)e^{-\frac{2\pi}{\sqrt{|\Delta|}}\rho^{-1}}\right), \tag{5.10}$$

on using (5.5). Formula (5.10) is the corrected form of Koshlyakov's [KoshI, (4.9)].

Finally, in the real quadratic case, Theorem 2.1 reads

$$\begin{aligned}&\sqrt{\rho}\left\{4\zeta_\Omega'(0)-4\sum_{k=1}^{\infty}F(k)K_0\left(\frac{2\pi k}{\sqrt{\Delta}}\rho\right)\right\}\\&=\frac{1}{\sqrt{\rho}}\left\{4\zeta_\Omega'(0)-4\sum_{k=1}^{\infty}F(k)K_0\left(\frac{2\pi i}{\sqrt{\Delta}}\frac{1}{\rho}\right)\right\}.\end{aligned} \tag{5.11}$$

Formula (5.11) falls under the $G_{0,2}^{2,0}\leftrightarrow G_{0,2}^{2,0}$ formula [MR, §3.1.2] or [MR, §9.2] cf. (9.44). A more general case has been treated as Example 3 in [BeI] and a more general form of (5.11) for general r_1 is given on [BeI, p.358].

Indeed, in the case $r_2=0\,(r_1\geq 2)$, we have

$$\operatorname{Res}\left(\Gamma\left(\frac{s}{2}\right)^{r_1}A^{-s}\zeta_\Omega(s)\,z^{-s},s=0\right)=2^{r_1}\frac{\zeta^{(r_1-1)}(0)}{(r_1-1)!}$$

(in [BeI], $\operatorname{Res}\left(\Gamma(s)^{r_1}A^{-2s}\zeta_\Omega(s)\,z^{-s},s=0\right)=2^{r_1-1}\frac{\zeta^{(r_1-1)}(0)}{(r_1-1)!}$), and

$$\begin{aligned}&\operatorname{Res}\left(\Gamma\left(\frac{s}{2}\right)^{r_1}A^{-s}\zeta_\Omega(s)\,z^{-s},s=1\right)\\&=\operatorname{Res}\left(\Gamma\left(\frac{1-s}{2}\right)^{r_1}A^{-(1-s)}\,\zeta_\Omega(1-s)\,z^{-s},s=1\right)\\&=\operatorname{Res}\Bigg(\left(\frac{-2}{s-1}\right)^{r_1}\Gamma\left(\frac{3}{2}-\frac{s}{2}\right)^{r_1}A^{-(1-s)}\\&\qquad\left(\cdots+\frac{\zeta_\Omega^{(r_1-1)}(0)}{(r_1-1)!}(1-s)^{r_1-1}+\cdots\right)z^{-s},s=1\Bigg)\\&=-2^{r_1}\frac{1}{(r_1-1)!}\,\zeta_\Omega^{(r_1-1)}(0)\,z^{-1}\end{aligned}$$

or

$$\operatorname{Res}\left(\Gamma\left(\frac{s}{2}\right)^{r_1}A^{-s}\zeta_\Omega(s)\,z^{-s},s=1\right)=\Gamma\left(\frac{1}{2}\right)^{r_1}A^{-1}\lambda h z^{-1},$$

where λh is the residue of $\zeta_\Omega(s)$ at $s=1$ given by (2.4). It may be also expressed in the case $\varkappa\leq 2$ as

$$\lambda h=-\frac{2^{r+1}\pi^{r_2}\zeta_\Omega^{(r)}(s)}{\sqrt{|\Delta|}}. \tag{5.12}$$

However, on [BeI, p.358] the residual function looks different. This is because he treats $A^{-2s}\zeta_\Omega(2s)$ as the Dirichlet series $\phi(s)$, so that $\chi(s) = \Gamma(s)^{r_1}\phi(s)z^{-s}$, with residue $A^{-1}\lambda h z^{-\frac{1}{2}}$, and he replaces z/A^2 by y, whence $z^{-\frac{1}{2}} = Ay^{-\frac{1}{2}}$, and so

$$\mathrm{Res}\left(\Gamma(s)^{r_1}A^{-2s}\zeta_\Omega(2s)\,z^{-s}, s = \frac{1}{2}\right) = \Gamma\left(\frac{1}{2}\right)^{r_1}\frac{1}{2}\lambda h y^{-\frac{1}{2}}.$$

Because of this replacement, the modular relation leads to a slightly different form ([BeI, p.358]). For this cf. [KTY7, §7].

5.2. *The* $H_{0,1}^{1,0} \leftrightarrow H_{0,1}^{1,0}$ *formula*

The special case of Theorem 1.1 with $M = 1$ includes the rational and imaginary quadratic cases of Theorem 5.1:

Theorem 5.2. *The functional equation*

$$\Gamma(d_1 + D_1 s)\varphi(s) = \Gamma(e_1 + E_1(r-s))\psi(r-s), \tag{5.13}$$

is equivalent to the modular relation

$$z^s\,\mathrm{X}(z,s\,|\,\Delta)$$

$$= \begin{cases} \displaystyle\sum_{k=1}^{\infty}\frac{\alpha_k}{\lambda_k^s}H_{0,1}^{1,0}\left(z\lambda_k \,\middle|\, \begin{matrix} - \\ (d_1+D_1 s, D_1)\end{matrix}\right) \\ \qquad\qquad \textit{if } L_1(s) \textit{ can be taken to the right of } \max\limits_{1\le h\le H}(\sigma_{\varphi_h}) \\ \displaystyle\sum_{k=1}^{\infty}\frac{\beta_k}{\mu_k^{r-s}}H_{0,1}^{1,0}\left(\frac{\mu_k}{z} \,\middle|\, \begin{matrix} - \\ (e_1+E_1(r-s), E_1)\end{matrix}\right) \\ \displaystyle\quad+\sum_{k=1}^{L}\mathrm{Res}\left(\chi(w)\,z^{s-w}, w = s_k\right) \\ \qquad\qquad \textit{if } L_2(s) \textit{ can be taken to the left of } \min\limits_{1\le i\le I}(r-\sigma_{\psi_i}) \end{cases} \tag{5.14}$$

or

$$\frac{z^{\frac{d_1+D_1 s}{D_1}}}{D_1}\sum_{k=1}^{\infty}\alpha_k\lambda_k^{\frac{d_1}{D_1}}e^{-(z\lambda_k)^{\frac{1}{D_1}}} = \frac{\left(z^{-1}\right)^{\frac{e_1+E_1(r-s)}{E_1}}}{E_1}\sum_{k=1}^{\infty}\beta_k\mu_k^{\frac{e_1}{E_1}}e^{-\left(\frac{\mu_k}{z}\right)^{\frac{1}{E_1}}} \tag{5.15}$$

$$+\sum_{k=1}^{L} \operatorname{Res}\Big(\chi(w)\, z^{s-w}, w = s_k\Big).$$

Proof follows from

$$\frac{1}{D_1}\sum_{k=1}^{\infty}\frac{\alpha_k}{\lambda_k^s}(z\lambda_k)^{\frac{d_1+D_1 s}{D_1}}e^{-(z\lambda_k)^{\frac{1}{D_1}}} = \frac{1}{E_1}\sum_{k=1}^{\infty}\frac{\beta_k}{\mu_k^{r-s}}\left(\frac{\mu_k}{z}\right)^{\frac{e_1+E_1(r-s)}{E_1}} e^{-\left(\frac{\mu_k}{z}\right)^{\frac{1}{E_1}}} \tag{5.16}$$

$$+\sum_{k=1}^{L} \operatorname{Res}\Big(\chi(w)\, z^{s-w}, w = s_k\Big).$$

Theorem 5.2 with no perturbation $d_1 = e_1 = 0$ is a typical example of the use of gamma transform (3.17).

Remark 5.2. It may remarked that the K-Bessel function $K_s(z)$ defined by (3.6) and its reduction to exponential functions (3.7) and (3.8) appeared in various contexts in relation to the lattice point problem. It was [HarMB] who first used the integral (3.6), without noticing that it is a K-Bessel function, in his research on the Epstein zeta-function, as has been elucidated by us [CZS3]. The following papers were not referred to there and we state them here.

(i) Bellman's papers. Indeed, [Bel3, (3)] is (3.8) with plus sign while [Siegl] is (3.8) with minus sign. We note the following chain of different symbols for the K-Bessel function, the first equality being [Bel5, (4.1)]

$$V_a(x,y) = x^{2a-1}V_a(|xy|), \quad V_a(|xy|) = \pi^a W_{2-a}((\pi xy)^2) \tag{5.17}$$

and

$$W_a(z) = \int_0^{\infty} e^{-zv-\frac{1}{v}} v^{-a}\, dv = 2z^{\frac{a-1}{2}}K_{a-1}(2\sqrt{z}), \quad \operatorname{Re} z > 0. \tag{5.18}$$

(ii) Siegel's proof of Hamburger's theorem. We notice that the main ingredient of Siegel's proof of Hamburger's theorem is indeed the partial fraction expansion for the cotangent function in [Vista I, Chapter 4]: The Riemann type functional equation

$$\Gamma\left(\frac{1}{2}s\right)\varphi(s) = \Gamma\left(\frac{1}{2}(r-s)\right)\psi(r-s) \tag{5.19}$$

with a simple pole at $s = r > 0$ with residue ρ is equivalent to

$$2\Gamma\left(\frac{1}{2}s\right)\sum_{k=1}^{\infty}\frac{\alpha_k}{(\lambda_k^2+z^2)^{\frac{1}{2}s}}$$
$$= 4z^{\frac{1}{2}(r-s)}\sum_{k=1}^{\infty}\frac{\beta_k}{\mu_k^{\frac{1}{2}(r-s)}}K_{\frac{1}{2}(r-s)}(2\mu_k z) + 2\Gamma\left(\frac{1}{2}s\right)\varphi(0)\,z^{-s}$$
$$+\rho\Gamma\left(\frac{1}{2}(s-r)\right)z^{r-s}. \tag{5.20}$$

This reduces for $s = 2, r = 1$ to the partial fraction expansion for a generalized cotangent function

$$2\sum_{k=1}^{\infty}\frac{\alpha_k}{\lambda_k^2+z^2} = 2\sqrt{\pi}z^{-1}\sum_{k=1}^{\infty}\beta_k e^{-2\mu_k z} + 2\varphi(0)z^{-2} + \rho\sqrt{\pi}z^{-1}, \tag{5.21}$$

which amounts to [Siegl, (10)].

Remarkably enough, Siegel deduces (5.21) not as the Fourier-Bessel expansion but as a consequence of the modular relation, a special case of (5.15):

$$2z^2\sum_{k=1}^{\infty}\alpha_k e^{-(z\lambda_k)^2} = 2z^{-(r-s)}\sum_{k=1}^{\infty}\beta_k e^{-\left(\frac{\mu_k}{z}\right)^2} + \mathrm{P}(z). \tag{5.22}$$

By a trivial change of variables and specifications, (5.22) amounts to [Siegl, 1.2,p.156]

$$2z^2\sum_{k=1}^{\infty}\alpha_k e^{-\pi n^2 x} = \frac{2}{\sqrt{x}}\sum_{k=1}^{\infty}\beta_k e^{-\frac{\pi n^2}{x}} + \mathrm{P}(x), \quad x > 0. \tag{5.23}$$

Then multiplying (5.23) by $e^{-\pi t^x}$ and integrating in x over $(0,\infty)$, thereby incorporating (3.8) with minus sign $K_{-\frac{1}{2}}$, thus uplifting $H_{0,1}^{1,0} \leftrightarrow H_{0,1}^{1,0}$ to $H_{1,1}^{1,1} \leftrightarrow H_{0,2}^{2,0}$.

(iii) Chowla and Selberg [CS], [SC] leaves the function undefined, denoting it e.g. by

$$I_a = \int_0^{\infty} e^{-\frac{\pi\sqrt{\Delta}n}{2a}(y+y^{-1})}y^{s-\frac{3}{2}}\,\mathrm{d}y = 2K_{s-\frac{1}{2}}\left(\frac{\pi\sqrt{\Delta}n}{a}\right). \tag{5.24}$$

5.3. *Koshlyakov's Y-function*

Complementary to the X-function is the Y-function which is the Mellin factor of the K-function in the sense that ([KoshI, (5.8)])

$$\int_0^{\infty} X\left(\frac{a}{x}\right)Y(bx)\,\mathrm{d}x = \frac{1}{b}\,K(ab), \quad a > 0, b > 0. \tag{5.25}$$

Cf. Example 6.1 below for a proof.

The Y-function is defined by ([KoshI, (5.4)])

$$\begin{aligned} Y(x) &= Y_{r_1,r_2}(x) \\ &= \frac{1}{2\pi i}\int_{(c)} \frac{\pi}{2\sin\left(\frac{\pi}{2}s\right)\,\Gamma^{r_1}\left(\frac{2-s}{2}\right)\Gamma^{r_2}(2-s)}\,x^{-s}\,\mathrm{d}s \\ &= \frac{1}{2\pi i}\int_{(c)} \frac{\Gamma(s)\Gamma(1-s)\cos\left(\frac{\pi}{2}s\right)}{\Gamma^{r_1}\left(\frac{2-s}{2}\right)\Gamma^{r_2}(2-s)}\,x^{-s}\,\mathrm{d}s, \quad c>0,\ x>0, \end{aligned} \tag{5.26}$$

for $0<\sigma<\frac{r_2+3}{\varkappa(\varkappa-1)}$. Or after some transformation,

$$\begin{aligned} Y(x) &= Y_{r_1,r_2}(x) \\ &= H^{1,1}_{2,r+3}\left(x\,\middle|\,\begin{matrix}(0,1),(\frac12,\frac12)\\ (0,1),(\frac12,\frac12),(0,\frac12),\dots,(0,\frac12),(-1,1),\dots,(-1,1)\end{matrix}\right) \\ &= \frac12 H^{1,1}_{1,r+2}\left(e^{\frac{\pi}{2}i}x\,\middle|\,\begin{matrix}(0,1)\\ (0,1),(0,\frac12),\dots,(0,\frac12),(-1,1),\dots,(-1,1)\end{matrix}\right) \\ &+ \frac12 H^{1,1}_{1,r+2}\left(e^{-\frac{\pi}{2}i}x\,\middle|\,\begin{matrix}(0,1)\\ (0,1),(0,\frac12),\dots,(0,\frac12),(-1,1),\dots,(-1,1)\end{matrix}\right). \end{aligned} \tag{5.27}$$

Though contradicting to our claim that the processing gamma factor is trivial, we may view Y as the case where the processing gamma factor ((1.4)) is $\Gamma(s\,|\,\Delta)=\frac{\Gamma(1-s)}{\Gamma^{r_1}(\frac{2-s}{2})\Gamma^{r_2}(2-s)}$ and the functional equation (2.7) is the case of the imaginary quadratic field. This fact will show its effect in due course.

We distinguish three cases. In the case $r_1=1, r_2=0$,

$$Y_{1,0}(x)=\frac{1}{2\pi i}\int_{(c)}\Gamma\left(\frac{s}{2}\right)x^{-s}\,\frac{\mathrm{d}s}{2}=G^{1,0}_{0,1}\left(x^2\,\middle|\,\begin{matrix}-\\0\end{matrix}\right)=e^{-x^2}. \tag{5.28}$$

In the case $r_1=0, r_2=1$, by (5.27),

$$Y_{0,1}(x)=\frac12\left(G^{1,1}_{1,2}\left(iz\,\middle|\,\begin{matrix}0\\0,-1\end{matrix}\right)+G^{1,1}_{1,2}\left(-iz\,\middle|\,\begin{matrix}0\\0,-1\end{matrix}\right)\right). \tag{5.29}$$

By (3.21),

$$G^{1,1}_{1,2}\left(z\,\middle|\,\begin{matrix}0\\s,-1\end{matrix}\right)=z^s\frac{\Gamma(1+s)}{\Gamma(2+s)}\,{}_1F_1\left(\begin{matrix}1+s\\2+s\end{matrix};-z\right).$$

By (3.33), this reduces further to

$$G^{1,1}_{1,2}\left(z\,\middle|\,\begin{matrix}0\\s,-1\end{matrix}\right)=z^s\frac{\Gamma(1+s)}{\Gamma(2+s)}(s+1)z^{-(s+1)}\gamma(s+1,z)=z^{-1}\gamma(s+1,z). \tag{5.30}$$

In the special case $s=0, r=1$, (5.30) reduces to

$$G_{1,2}^{1,1}\left(z \,\middle|\, \begin{matrix} 0 \\ 0,-1 \end{matrix}\right) = z^{-1}\gamma(1,z) = \frac{1-e^{-z}}{z}. \tag{5.31}$$

Hence

$$Y_{0,1}(x) = \frac{\sin(x)}{x} = \mathrm{si}(x), \tag{5.32}$$

the sinus cardinalis function which plays an important role in signal transmission (cf. [Spl]). Below we will give a modular relation version of the sampling theorem in Theorem 5.4.

Finally, in the case of

$$Y_{2,0}(x) = \frac{1}{2\pi i}\int_{(c)} \frac{\pi}{2\sin\left(\frac{\pi}{2}s\right)\Gamma\left(1-\frac{s}{2}\right)^2} x^{-s}\mathrm{d}s,$$

we rewrite the integrand as $\frac{\Gamma\left(\frac{s}{2}\right)}{\Gamma\left(1-\frac{s}{2}\right)}\frac{x^{-s}}{2}$ and then express $Y_{2,0}(x)$ as the G-function:

$$Y_{2,0}(x) = \frac{1}{2}H_{0,2}^{1,0}\left(x \,\middle|\, \begin{matrix} - \\ (0,\frac{1}{2}),(0,\frac{1}{2}) \end{matrix}\right) = G_{0,2}^{1,0}\left(x^2 \,\middle|\, \begin{matrix} - \\ 0,0 \end{matrix}\right) = J_0(2x)$$

by (3.19), $J_s(x)$ being the Bessel function defined by (3.1).

We recall that the X- and Y-functions play an important role in the derivation of the Riemann-Siegel integral formula [KTY7, §4.2], in that a recourse to the formula ([KoshI, (9.10)])

$$\int_1^\infty X(ax)Y(bx)\,x\,\mathrm{d}x = \frac{Z(a,b)}{a^2+b^2}, \quad a>0, b>0$$

was essential.

In what follows we shall deduce Theorem 5.4 as a special case of the main theorem of modular relations. One of the forms of the main theorem in [MR] reads

Theorem 5.3. *With the gamma factor given by* (1.4) *with parameters* (1.3), *we have the modular relation*

$$\sum_{k=1}^\infty \frac{\alpha_k}{\lambda_k^s} H_{p,q+1}^{m+1,n}\left(z\lambda_k \,\middle|\, \begin{matrix} \{(1-a_j,A_j)\}_{j=1}^n, \{(a_j,A_j)\}_{j=n+1}^p \\ (s,1), \{(b_j,B_j)\}_{j=1}^m, \{(1-b_j,B_j)\}_{j=m+1}^q \end{matrix}\right) \tag{5.33}$$

$$= \sum_{k=1}^\infty \frac{\beta_k}{\mu_k^{r-s}} H_{q,p+1}^{n+1,m}\left(\frac{\mu_k}{z} \,\middle|\, \begin{matrix} \{(1-b_j,B_j)\}_{j=1}^m, \{(b_j,B_j)\}_{j=m+1}^q \\ (r-s,1), \{(a_j,A_j)\}_{j=1}^n, \{(1-a_j,A_j)\}_{j=n+1}^p \end{matrix}\right)$$

$$+ \mathrm{P}(z)$$

is equivalent to the Hecke type functional equation

$$\Gamma(s)\varphi(s) = \Gamma(r-s)\varphi(r-s), \tag{5.34}$$

where $\mathrm{P}(z) = \sum_{k=1}^{L} \mathrm{Res}\Big(\Gamma(w-s\,|\,\Delta)\,\chi(w)\,z^{s-w}, w = s_k\Big)$ *indicates the residual function.*

We assume that $\varphi(s)$ has a simple pole at $s=1$ with residue ρ_φ, so that $\chi(s) = \Gamma(s)\varphi(s)$ has simple poles at $s=1$ and $s=0$.

Formally we apply the case $m=0, n=1, p=2, q=2$ and $a_1 = 1, A_1 = 1, a_2 = \frac{1}{2}, A_2 = \frac{1}{2}$ $b_1 = -1, B_1 = 1, b_2 = \frac{1}{2}, B_2 = \frac{1}{2}$ of Theorem 5.3 with

$$\Gamma(w-s\,|\,\Delta) = \frac{\Gamma(1+s-w))}{\Gamma(2+s-w)\Gamma\left(\frac{1+s}{2}-\frac{1}{2}w\right)\Gamma\left(\frac{1-s}{2}+\frac{1}{2}w\right)}.$$

However, we appeal to (5.29) by putting

$$\Gamma(w-s\,|\,\Delta) = \frac{\Gamma(1-(w-s))}{\Gamma(2-(w-s))}\cos\frac{\pi}{2}(w-s)$$

and expressing the cosine function as the sum of exponential functions. Then this amounts to The residual function is computed to be

$$\begin{aligned}\mathrm{P}(z) &= \frac{\Gamma(1+s)}{\Gamma(2+s)}\cos\frac{\pi s}{2}z^s\varphi(0) + \frac{\Gamma(s)}{\Gamma(1+s)}z^{s-1}\sin\frac{\pi s}{2}\rho_\varphi \\ &= \frac{1}{s+1}\cos\frac{\pi s}{2}z^s\varphi(0) + \frac{1}{s}z^{s-1}\sin\frac{\pi s}{2}\rho_\varphi.\end{aligned} \tag{5.35}$$

The modular relation (5.33) amounts to the case $m=0, n=1, p=1, q=1$:

$$\begin{aligned}&\sum_{k=1}^{\infty}\frac{\alpha_k}{\lambda_k^s}\left(G_{1,2}^{1,1}\left(iz\lambda_k\,\middle|\,\begin{matrix}0\\ s,-1\end{matrix}\right) + G_{1,2}^{1,1}\left(-iz\lambda_k\,\middle|\,\begin{matrix}0\\ s,-1\end{matrix}\right)\right) \\ &= \sum_{k=1}^{\infty}\frac{\beta_k}{\mu_k^{r-s}}\left(G_{1,2}^{2,0}\left(i\frac{\mu_k}{z}\,\middle|\,\begin{matrix}2\\ r-s,1\end{matrix}\right) + G_{1,2}^{2,0}\left(-i\frac{\mu_k}{z}\,\middle|\,\begin{matrix}2\\ r-s,1\end{matrix}\right)\right) \\ &\quad + 2\mathrm{P}(z).\end{aligned} \tag{5.36}$$

The left-hand side function is (5.29), which is the sinus cardinalis function in (5.32).

On the right-hand side, by (3.29)

$$G_{1,2}^{2,0}\left(z\,\middle|\,\begin{matrix}2\\ r-s,1\end{matrix}\right) = e^{-z}\,z^{r-s}\,U\big(1, r-s, z\big).$$

By (3.34), this reduces further to

$$G_{1,2}^{2,0}\left(z\,\middle|\,\begin{matrix}2\\ r-s,1\end{matrix}\right)=e^{-z}\,z^{r-s}\,z^{1-(r-s)}e^{z}\Gamma(r-s-1,z)=z\Gamma(r-s,z). \tag{5.37}$$

In the special case $s=0, r=1$ of (5.36), since (5.37) reduces to

$$G_{1,2}^{2,0}\left(z\,\middle|\,\begin{matrix}2\\ 1,1\end{matrix}\right)=z\Gamma(1,z)=ze^{-z}, \tag{5.38}$$

we conclude that

$$\begin{aligned}\frac{1}{2}&\left(G_{1,2}^{2,0}\left(iz\,\middle|\,\begin{matrix}2\\ 1,1\end{matrix}\right)+G_{1,2}^{2,0}\left(-iz\,\middle|\,\begin{matrix}2\\ 1,1\end{matrix}\right)\right)\\ &=z\sin z.\end{aligned} \tag{5.39}$$

(5.35) reduces for $s=0$ to $\varphi(0)+\frac{\pi}{2}z^{-1}\rho_\varphi$. Thus we have proved

Theorem 5.4.

$$\sum_{k=1}^{\infty}\alpha_k\,\mathrm{si}(z\lambda_k)=\frac{1}{z}\sum_{k=1}^{\infty}\beta_k\sin\frac{\mu_k}{z}+\varphi(0)+\frac{\pi}{2}z^{-1}\rho_\varphi. \tag{5.40}$$

This is interesting in the light of the sampling theorem, i.e. the left-hand side of (5.40) may be thought of as the Fourier series of the function $f(z)$ in question and the right-hand side is the sampling series as given in the sampling theorem.

6. Modular relations in integral form

Under the heading "integration with respect to the parameter", Koshlyakov gave a number of intriguing integral identities, which may be viewed as modular relations in integral form. In §6.1, we give a typical example of this which is a generalization of Ramanujan's integral formula. Then in §6.3 we shall interpret this and many other integral identities as manifestations of the identity between Mellin transform and its inversion.

6.1. *Generalization of Ramanujan's integral formula*

For the present let ($a>0$)

$$\Psi(a)=\sqrt{a^3}\int_0^\infty xY_{r_1,r_2}(ax)\left(\sigma(x)+\frac{\zeta_\Omega(0)}{\pi}\frac{1}{x}\right)\mathrm{d}x \tag{6.1}$$

(according to [KoshII, (19.8),p.234] with $a=Ae^{-2z}$).

Theorem 6.1. ([KoshII, (19.9), p.234], [KoshIII, (31.4), p.308]) *Under the reciprocal relation*

$$ab = A^2, \tag{6.2}$$

the modular relation $\Psi(a) = \Psi(b)$ *holds true or*

$$\begin{aligned}\sqrt{a^3} \int_0^\infty x Y_{r_1,r_2}(ax)\left(\sigma(x) + \frac{\zeta_\Omega(0)}{\pi}\frac{1}{x}\right) \mathrm{d}x \\ = \sqrt{b^3} \int_0^\infty x Y_{r_1,r_2}(bx)\left(\sigma(x) + \frac{\zeta_\Omega(0)}{\pi}\frac{1}{x}\right) \mathrm{d}x.\end{aligned} \tag{6.3}$$

This is proved as a consequence of his summation formula (I). Here we follow another proof depending on the theory of σ-series. First we recall the integral expression for the σ-function (6.19) below with which the theory of Y-functions is intimately connected.

Corollary 6.1.

(i) *In the rational case,* (6.3) *leads to Ramanujan's formula*

$$\begin{aligned}\sqrt{a^3} \int_0^\infty x e^{-(ax)^2}\left(\frac{1}{e^{2\pi x}-1} - \frac{1}{2\pi x}\right) \mathrm{d}x \\ = \sqrt{b^3} \int_0^\infty x e^{-(bx)^2}\left(\frac{1}{e^{2\pi x}-1} - \frac{1}{2\pi x}\right) \mathrm{d}x,\end{aligned} \tag{6.4}$$

where (6.2) *amounts to*

$$ab = \pi. \tag{6.5}$$

(ii) *In the imaginary quadratic case,* (6.3) *amounts to the modular relation* (5.10) *in the equivalent form*

$$\sqrt{a}\left(h + w\sum_{k=1}^\infty F(k)e^{-na}\right) = \frac{1}{\sqrt{b}}\left(h + w\sum_{k=1}^\infty F(k)e^{-nb}\right), \tag{6.6}$$

which is a corrected form of [KoshIII, (31.7),p.308] and where (6.2) *amounts to*

$$ab = \frac{4\pi^2}{|\Delta|}. \tag{6.7}$$

(iii) *In the real quadratic case,* (6.3) *leads to the modular relation*

$$\sqrt{a^3} \int x J_0(2ax)\sigma(x)\,\mathrm{d}x = \sqrt{b^3} \int x J_0(2bx)\sigma(x)\,\mathrm{d}x, \tag{6.8}$$

where $\sigma(x) = \sigma_{0,2}(x)$ *indicates the basic series defined in* (6.19) *and* (6.2) *amounts to*

$$ab = \frac{\pi^2}{\Delta}. \tag{6.9}$$

Proof. (i) follows from

$$\sigma_{1,0}(x) = \frac{1}{e^{2\pi x} - 1}. \tag{6.10}$$

(ii) follows from

$$\sum_{n=1}^{\infty} F_\Omega(n) e^{-an} = -\frac{h}{w} + \frac{2\pi h}{w\sqrt{|\Delta|}} \frac{1}{a} + \int_0^\infty \sin(ax)\sigma_{0,2}(x)\, dx, \tag{6.11}$$

which is a corrected form of [KoshII, (8.16),p.231] (stated again on [KoshII, p.308]). This in turn follows from the generalized Plana summation formula (Theorem 7.1 below).

(iii) is a simple rewriting. □

In the case of the real quadratic field, it is Koshlyakov's [KoshIII, (31.10)] rather than (6.8) that leads to the modular relation.

Theorem 6.2. *The equality*

$$\int_0^\infty J_0(2ax)\sigma(x)\, dx = \int_0^\infty J_0(2bx)\sigma(x)\, dx, \tag{6.12}$$

valid under the reciprocal relation $ab = \frac{\pi^2}{\Delta}$ *leads to the modular relation* (5.11).

Proof. Recall that (6.20) below takes the form

$$\sigma(x) = \frac{1}{2\pi i} \int_{(c)} \left(\frac{\sqrt{\Delta}}{\pi}\right)^{1-s} \frac{\Gamma^2\left(\frac{1-s}{2}\right)}{\Gamma^2\left(\frac{s}{2}\right)} \frac{\pi\zeta_\Omega(1-s)}{2\sin\left(\frac{\pi}{2}s\right)} x^{s-1}\, ds, \quad 0 < c < 1. \tag{6.13}$$

We use (6.13) and the Mellin inversion formula for the J-Bessel function defined by (3.1):

$$\int_{(c)} J_0(2ax) x^s \frac{dx}{x} = \frac{\Gamma^2\left(\frac{s}{2}\right)\sin\left(\frac{\pi s}{2}\right)}{2\pi a^s} \quad 0 < \sigma < \frac{3}{2} \tag{6.14}$$

to deduce

$$\int_0^\infty J_0(2Ax)\sigma(x)\, dx = \frac{1}{2\pi i} \int_{(c)} \left(\frac{\sqrt{\Delta}}{\pi}\right)^{1-s} \Gamma^2\left(\frac{1-s}{2}\right) \zeta_\Omega(s) \frac{ds}{4\left(\frac{\sqrt{\Delta}a}{\pi}\right)^s}. \tag{6.15}$$

Applying the functional equation (2.3), we conclude (6.12).

To deduce (5.11) from (6.12) we need one more formula which follows from a form of the Plana summation formula:

$$\sum_{k=1}^{\infty} F(k)K_0(2k) = -\frac{\pi\zeta_\Omega'(0)}{\sqrt{\Delta}a} + \pi\int_0^\infty J_0(2bx)\sigma(x)\,\mathrm{d}x. \tag{6.16}$$

□

6.2. *Koshlyakov's σ-series*

Recall the functional equation (2.7) for the Dedekind zeta-function $\zeta_\Omega(s)$ of a number field Ω. Let $G(s) = G_\Omega(s)$ denote the quotient of gamma factors:

$$G(1-s) = \frac{\Gamma^{r_1}\left(\frac{s}{2}\right)\Gamma^{r_2}(s)}{\Gamma^{r_1}\left(\frac{1-s}{2}\right)\Gamma^{r_2}(1-s)} \tag{6.17}$$

and consider the kernel, which we call *Koshlyakov's K-function*

$$K(x) = K_{r_1,r_2}(x) = \frac{1}{2\pi i}\int_{(c)} \frac{\pi}{2\cos\left(\frac{\pi}{2}s\right)}\frac{G(1-s)}{x^s}\mathrm{d}s \tag{6.18}$$

for $c > 0$ and $x > 0$ ($\mathrm{Re}(x) > 0$ being allowed).

Then Koshlyakov considers the basic *σ-series*

$$\sigma(x) = \sigma_{r_1,r_2}(x) = \frac{A}{\pi}\sum_{n=1}^{\infty} F(n)K_{r_1,r_2}(A^2xn) \tag{6.19}$$

$x > 0\,(\mathrm{Re}(x) > 0)$.

Koshlyakov transforms the series (6.19) by Cauchy's residue theorem into [KoshI, (6.12),p.124, (7.4),p.125], [KoshIII, (31.1),p.307]

$$\begin{aligned}\sigma(x) + \frac{\zeta_\Omega(0)}{\pi x} &= \frac{1}{2\pi i}\int_{(c)} \frac{A^{1-2s}G(1-s)\zeta_\Omega(s)}{2\cos\left(\frac{\pi}{2}s\right)}\frac{\mathrm{d}s}{x^s} \\ &= \frac{1}{2\pi i}\int_{(c)} \frac{A^{2s-1}G(s)\zeta_\Omega(1-s)}{2\sin\left(\frac{\pi}{2}s\right)}\frac{\mathrm{d}s}{x^s}, \quad 0 < c < 1.\end{aligned} \tag{6.20}$$

Proof of Theorem 6.1. On the basis of (6.20), we may prove our theorem. Equivalent to (5.26) ([KoshIII, (31.2),p.307]) is

$$\int_0^\infty Y_{r_1,r_2}(ax)x^s\,\mathrm{d}x = \frac{\pi}{2\cos\frac{\pi s}{2}\Gamma^{r_1}\left(\frac{1-s}{2}\right)\Gamma^{r_2}(1-s)\,a^{s+1}}, \tag{6.21}$$

for $a > 0$, $-1 < \sigma < \frac{r_2+3}{\varkappa(\varkappa-1)} - 1$.

Multiplying (6.20) by $xY_{r_1,r_2}(ax)$ and *integrating in the parameter* x over $(0,\infty)$, we obtain

$$\frac{1}{\sqrt{a^3}}\Psi(a) = \int_0^\infty xY_{r_1,r_2}(ax)\left(\sigma(x)+\frac{\zeta_\Omega(0)}{\pi x}\right)\,\mathrm{d}x \tag{6.22}$$
$$= \frac{1}{2\pi i}\int_{(c)}\frac{\pi A^{2s-1}G(s)\zeta_\Omega(1-s)}{2\sin\pi s\Gamma^{r_1}\left(\frac{1-s}{2}\right)\Gamma^{r_2}(1-s)}\int_0^\infty Y_{r_1,r_2}(ax)x^s\,\mathrm{d}x\frac{\mathrm{d}s}{a^{s+1}}$$

on changing the order of integration. Hence incorporating (6.21), we conclude [KoshIII, (31.3),p.307]:

$$\Psi(a) = \frac{\sqrt{a^3}}{2\pi i}\int_{(c)}\frac{\pi A^{2s-1}\zeta_\Omega(1-s)}{2\sin(\pi s)\,\Gamma^{r_1}\left(\frac{s}{2}\right)\Gamma^{r_2}(s)}\frac{\mathrm{d}s}{a^{s+1}}. \tag{6.23}$$

Note that using (6.17), we may write (2.7) as

$$A^{2s-1}\zeta_\Omega(1-s) = G(1-s)\zeta_\Omega(s).$$

Hence, (6.23) may be written as

$$\Psi(a) = \frac{\sqrt{a^3}}{2\pi i}\int_{(c)}\frac{\pi\zeta_\Omega(s)}{2\sin(\pi s)\,\Gamma^{r_1}\left(\frac{1-s}{2}\right)\Gamma^{r_2}(1-s)}\frac{\mathrm{d}s}{a^{s+1}} \tag{6.24}$$
$$= \frac{A^3}{\sqrt{a^3}}\frac{1}{2\pi i}\int_{(c)}\frac{\pi\zeta_\Omega(1-s)}{2\sin(\pi s)\,\Gamma^{r_1}\left(\frac{s}{2}\right)\Gamma^{r_2}(s)}\frac{\mathrm{d}s}{\left(\frac{A^2}{a}\right)^{s+1}},$$

which leads to (6.3) on putting $b=\frac{A^2}{a}$. Hence (6.2) also holds true, completing the proof.

Proof of Corollary 6.1 (ii) (6.3) leads to

$$\sqrt{a}\int_0^\infty \sin ax\left(\sigma(x)-\frac{h}{w}\frac{1}{\pi x}\right)\,\mathrm{d}x \tag{6.25}$$
$$= \sqrt{b}\int_0^\infty \sin bx\left(\sigma(x)-\frac{h}{w}\frac{1}{\pi x}\right)\,\mathrm{d}x,$$

which is a corrected form of [KoshIII, (31.6), p.308]. To transform (6.25) into the modular relation we need some more data. One is

$$\sum_{k=1}^\infty F(k)e^{-ak} = -\frac{h}{w}+\frac{h}{w}\frac{2\pi}{\sqrt{|\Delta|}}\frac{1}{a}+\int_0^\infty \sin(ax)\sigma(x)\,\mathrm{d}x, \tag{6.26}$$

which is a corrected form of the formulas squeezed by [KoshIII, (31.6), p.308] and [KoshIII, (31.7), p.308]. This is also proved by the summation formula (I) and the correct form of the formula is the first formula on [KoshIII, p.256]

$$\sum_{k=1}^{\infty} F(k)e^{-ak} = \zeta_\Omega(0) - \frac{2\pi}{\sqrt{|\Delta|}} \frac{\zeta_\Omega(0)}{a} + 2\int_0^\infty \sin(ax)\sigma(x)\,\mathrm{d}x, \tag{6.27}$$

and accordingly, [KoshII, (18.6), p.231] should read

$$\sum_{k=1}^{\infty} F(k)e^{-A\rho k} = \zeta_\Omega(0)f(0) - \frac{\zeta_\Omega(0)}{\rho} + 2\int_0^\infty \sin(A\rho x)\sigma(x)\,\mathrm{d}x, \tag{6.28}$$

where $\zeta_\Omega(0) = -\frac{h}{w}$ and A is given by (2.1). The other is a well-known evaluation for the sinus cardinalis function

$$\int_0^\infty \frac{\sin x}{x}\,\mathrm{d}x = \frac{\pi}{2}. \tag{6.29}$$

Appealing to (6.2), we deduce (6.8).

6.3. *Integration in the parameter*

Theorem 6.3. *Suppose the inverse Mellin transform*

$$f(x) = \frac{1}{2\pi i}\int_{(c)} \mathfrak{F}(s)x^{-s}\,\mathrm{d}s \quad c > 0 \tag{6.30}$$

and the Mellin transform

$$b^{-s}\mathfrak{G}(s) = \int_0^\infty g(bx)x^s \frac{\mathrm{d}x}{x} \quad \sigma > 0,\ b > 0 \tag{6.31}$$

are given. Then we have the integral formulas

$$\int_0^\infty f\left(\frac{a}{x}\right) g(bx)\,\mathrm{d}x = b^{-1}\frac{1}{2\pi i}\int_{(c)} \mathfrak{F}(s)\mathfrak{G}(1+s)(ab)^{-s}\,\mathrm{d}s, \quad a > 0, \tag{6.32}$$

and

$$\int_0^\infty f(ax)g(bx)\,\mathrm{d}x = b^{-1}\frac{1}{2\pi i}\int_{(c)} \mathfrak{F}(s)\mathfrak{G}(1-s)\left(\frac{a}{b}\right)^{-s}\mathrm{d}s, \quad a > 0, \tag{6.33}$$

provided that the following growth conditions are satisfied: Writing $\tilde{f}(x,t,c) = \mathfrak{F}(c+it)x^{c+it}$, we should have

$$\int_{x_0}^\infty \tilde{f}(x,t,c)g(bx)\,\mathrm{d}x << \phi_1(t), \quad \int_{t_0}^\infty \tilde{f}(x,t,c)g(bx)\,\mathrm{d}t << \phi_2(x), \tag{6.34}$$

for any t_0, x_0 big enough and $\phi_j(t) << e^{-q_j t}$, $q_j > 0$.

Indeed, substituting (6.30) in the left-hand side of (6.32), the left-hand side becomes

$$\begin{aligned}&\int_0^\infty g(bx)\left(\frac{1}{2\pi i}\int_{(c)}\mathfrak{F}(s)(ax)^{-s}\,\mathrm{d}s\right)\frac{\mathrm{d}x}{x}\\&=\frac{1}{2\pi i}\int_{(c)}\mathfrak{F}(s)a^{-s}\,\mathrm{d}s\left(\int_0^\infty g(bx)x^{-s}\,\mathrm{d}x\right)\end{aligned}\tag{6.35}$$

since the interchange of the repeated infinite integrals is permissible by absolute convergence. The inner integral is $b^{s-1}\mathfrak{G}(1-s)$ by the change of variables.

Recall the X-function defined by (2.9) and the Y-function by (5.26).

Example 6.1. The X-function and the Y-function are the Mellin factor of the K-function in the sense that ([KoshI, (5.8)])

$$\int_0^\infty X\left(\frac{a}{x}\right)Y(bx)\,\mathrm{d}x=\frac{1}{b}\,K(ab),\quad a>0,b>0.\tag{6.36}$$

Indeed, in this case

$$\mathfrak{F}(s)\mathfrak{G}(s+1)=\frac{\pi\Gamma^{r_1}\left(\frac{s}{2}\right)\Gamma^{r_2}(s)}{2\cos\left(\frac{\pi}{2}s\right)\,\Gamma^{r_1}\left(\frac{1-s}{2}\right)\Gamma^{r_2}(1-s)}=\pi\frac{G(1-s)}{2\cos\left(\frac{\pi}{2}s\right)}.$$

Example 6.2. If we define the X-function ([KoshI, p.121]) by

$$\frac{1}{2\pi i}\int_0^\infty X_{r_1,r_2}(x)\,\mathrm{d}s,=\Gamma^{r_1}\left(\frac{s}{2}\right)\Gamma^{r_2}(s)x^{-s}\quad c>0,\quad \sigma>0,\tag{6.37}$$

for $x>0$ ($\mathrm{Re}(x)>0$) and the Y-function by ([KoshI, (5.4)])

$$\begin{aligned}&Y(x)=Y_{r_1,r_2}(x)\\&=\frac{1}{2\pi i}\int_{(c)}\frac{\pi}{2\sin\left(\frac{\pi}{2}s\right)\,\Gamma^{r_1}\left(\frac{2-s}{2}\right)\Gamma^{r_2}(2-s)}x^{-s}\,\mathrm{d}s,\quad c>0,\,x>0.\end{aligned}\tag{6.38}$$

Then we again have (6.36).

From (5.26) and (6.20) we deduce [KoshI, (31.3)]

$$\begin{aligned}&\int_0^\infty xY_{r_1.r_2}\left(\sigma(x)+\frac{\zeta_\Omega(0)}{x}\right)\mathrm{d}x\\&=\frac{1}{2\pi i}\int_{(c)}\frac{\pi A^{2s-1}G(s)\zeta_\Omega(1-s)}{2\sin(\pi s)\Gamma^{r_1}\left(\frac{s}{2}\right)\Gamma^{r_2}(s)}a^{-s-1}\,\mathrm{d}s,\quad 0<c<1.\end{aligned}\tag{6.39}$$

Applying the functional equation

$$\zeta_\Omega(1-s)=A^{2s-1}G(1-s)\zeta_\Omega(s)\tag{6.40}$$

and then substituting $s = 1 - z$, we transform (6.39) into

$$\begin{aligned} &\frac{1}{2\pi i}\int_{(c)} \frac{\pi\zeta_\Omega(s)}{2\sin(\pi s)\,\Gamma^{r_1}\left(\frac{1-s}{2}\right)\Gamma^{r_2}(1-s)}\, a^{-s-1}\,\mathrm{d}s \\ &= \frac{A^3}{a^3}\frac{1}{2\pi i}\int_{(c_1)} \frac{\pi A^{2s-1}G(s)\zeta_\Omega(1-s)}{2\sin(\pi s)\,\Gamma^{r_1}\left(\frac{z}{2}\right)\Gamma^{r_2}(z)}\, a^{-s-1}\,\mathrm{d}z, \quad 0 < c_1 < 1, \end{aligned} \tag{6.41}$$

which is the corrected form of [KoshI, ll.4-5, p.308].

Example 6.3. Koshlyakov's [KoshI, (12.19)] is the inversion of the formula in Lemma 7.1

$$\frac{f(ix) - f(-ix)}{2} = \frac{1}{2\pi i}\int_{(c)} \mathfrak{F}(s)\sin\left(\frac{\pi s}{2}\right) x^{-s}\,\mathrm{d}s. \tag{6.42}$$

Koshlyakov proved [KoshI, (6.2)]

$$\zeta_\Omega(s) = 2\sin\left(\frac{\pi s}{2}\right)\int_0^\infty \sigma(x)x^{-s}\,\mathrm{d}x. \tag{6.43}$$

From these we conclude [KoshI, l.5,p.217]

$$-2\int_0^\infty \frac{f(ix) - f(-ix)}{2}\sigma(x)x^{-s}\,\mathrm{d}x = \frac{1}{2\pi i}\int_{(c)} \mathfrak{F}(s)\zeta_\Omega(s)\,\mathrm{d}s. \tag{6.44}$$

7. Generalization of the Plana summation formula

The following lemma gives a possibility of continuing the resulting integral to the left.

Lemma 7.1. *Suppose that $f(x)$ is defined and analytic in the half-plane* $\mathrm{Re}\, x \geq 0$. *Then the two representations for $\mathfrak{F}(s)$ (inversion of* (6.30)*) and*

$$\mathfrak{F}(s) = -\frac{1}{\sin\frac{\pi}{2}s}\int_0^\infty \frac{f(ix) - f(-ix)}{2i} x^s \frac{\mathrm{d}x}{x} \tag{7.1}$$

for $0 < \sigma < 2$, are equivalent. Further, in the case where

$$f(ix) - f(-ix) = o(x), \quad \sigma \to 0+, \tag{7.2}$$

(7.1) *holds true for $-1 < \sigma < 0$.*

Proof. By rotating the positive real axis by $\frac{\pi}{2}$, we obtain

$$\mathfrak{F}(s) = \int_0^\infty f(ix)(ix)^{s-1}\, i\mathrm{d}x = e^{\frac{\pi i}{2}s}\int_0^\infty f(ix)\, x^{s-1}\mathrm{d}x$$

and similarly

$$\mathfrak{F}(s) = e^{-\frac{\pi i}{2}s}\int_0^\infty f(-ix)\, x^{s-1}\mathrm{d}x.$$

Hence

$$\sin \frac{\pi s}{2}\mathfrak{F}(s) = \frac{e^{\frac{\pi i}{2}s} - e^{-\frac{\pi i}{2}s}}{2i}\mathfrak{F}(s) = -\int_0^\infty \frac{f(ix) - f(-ix)}{2i}x^{s-1}\mathrm{d}x,$$

which is (7.1). □

The following theorem is a slight generalization of [KoshII, p. 217] which is the case $z = 1$.

We refer to the following growth condition of Stirling type:

$$\mathfrak{F}(s) = O\left(e^{-\frac{\pi}{2}|t|}|t|^{a\sigma - b}\right) \tag{7.3}$$

uniformly in σ for $\sigma_1 \le \sigma \le \sigma_2$, where $a > 0, b > 0$ are subject to

$$\frac{a}{2} + b \ge 1. \tag{7.4}$$

Theorem 7.1. (Generalized Plana summation formula) *Suppose that the Mellin transform $\mathfrak{F}(s)$ of the function $f(x)$ $(x > 0)$ satisfies the growth condition (7.3). Suppose $f(x)$ is defined and analytic in the half-plane $\mathrm{Re}\, x \ge 0$ in Lemma 7.1, that (7.2) holds, and that $\mathfrak{F}(s)$ is regular in the strip $\beta \le \sigma \le \alpha$ $(-1 < \beta < 1,\ 0 < \alpha)$ except for a simple pole at $s = 0$. Then we have Plana's summation formula as a modular relation ($\mathrm{Re}\, z > 0$)*

$$\begin{aligned}\sum_{n=1}^{\infty} F(n) f(nz) &= \zeta_\Omega(0) \operatorname*{Res}_{s=0} \mathfrak{F}(s) \\ &\quad + \mathfrak{F}(1) z^{-1} - \frac{2}{z}\int_0^\infty \frac{f(ix) - f(-ix)}{2i}\sigma\left(\frac{x}{z}\right)\mathrm{d}x,\end{aligned} \tag{7.5}$$

provided that the infinite series and the integral in (7.5) are (absolutely) convergent. The first term on the right may be expressed as $\zeta_\Omega(0)f(0)$ under some growth condition on $\mathfrak{F}(s)$ and the second term is $\frac{\rho_\Omega}{z}\int_0^\infty f(x)\mathrm{d}x$, where $\rho_\Omega = -\frac{2^{r+1}\pi^{r_2}\zeta_\Omega^{(r)}(s)}{\sqrt{|\Delta|}}$ indicates the residue of the Dedekind zeta-function at $s = 1$ given in (5.12).

Proof of Theorem 7.1. We move the line of integration to $\sigma = -\alpha$ by the Cauchy residue theorem. Then we have for $\mathrm{Re}\, z > 0$ and $\alpha > 1$, using the estimates

$$\zeta_\Omega(s) = O\left(|t|^{\varkappa(1-\sigma)}\right), \quad t \le 0 \tag{7.6}$$

and (7.3)

$$\sum_{n=1}^{\infty} F(n)f(nz) = \frac{1}{2\pi i}\int_{(\alpha)} \mathfrak{F}(s)\zeta_\Omega(s)z^{-s}\,\mathrm{d}s \tag{7.7}$$
$$= R_0 + R_1 + \frac{1}{2\pi i}\int_{(1+\alpha)} \mathfrak{F}(1-s)\zeta_\Omega(1-s)z^{s-1}\,\mathrm{d}s,$$

by (7.3), where $\alpha > 1$ and we made the change of variable s by $1-s$ in the integral on the right, and where R_j indicates the residue at $s = j$, $j = 0, 1$ of the integrand, so that

$$R_1 = \rho_\Omega \mathfrak{F}(1)z^{-1}, \quad R_0 = -\zeta_\Omega(0)\operatorname*{Res}_{s=0}\mathfrak{F}(s).$$

We apply the functional equation (2.7) in the form (cf. (6.17))

$$\zeta_\Omega\,(1-s) = A^{1-2s}G(1-s)\zeta_\Omega\,(s) \tag{7.8}$$

so that the integrand of the integral, say I_α, on the right of (7.7) reduces, we see that

$$I_\alpha = \frac{A}{2\pi i z}\int_{(\alpha)} \mathfrak{F}(1-s)G(1-s)\sum_{n=1}^{\infty} F(n)\left(\frac{A^2 n}{z}\right)^{-s}\mathrm{d}s. \tag{7.9}$$

Now we use Lemma 7.1 in the form ($-1 < \sigma < 0$)

$$\mathfrak{F}(1-s) = -\frac{1}{\cos\frac{\pi}{2}s}\int_0^\infty \frac{f(ix)-f(-ix)}{2i}x^{-s}\,\mathrm{d}x \tag{7.10}$$

and change the order of integration in (7.9) (which is legitimate by the absolute convergence as is proved in [KoshII]), we obtain

$$I_\alpha = \frac{2}{z}\frac{A}{\pi}\int_0^\infty \frac{f(ix)-f(-ix)}{2i}\sum_{n=1}^{\infty}\sigma_1(x,n)\,\frac{\mathrm{d}x}{x}, \tag{7.11}$$

where

$$\sigma_1(x,n) = \frac{1}{2\pi i}\int_{(c)} \frac{\pi}{2\cos(\frac{\pi}{2}s)}G(1-s)(A^2nx/z)^{-s}\mathrm{d}s = K(A^2nx/z), \tag{7.12}$$

whence $\frac{A}{\pi}\sum_{n=1}^{\infty}\sigma_1(x,n) = \sigma\left(\frac{x}{z}\right)$ and (7.7) follows, completing the proof.

8. Integrated modular relations

In this section we shall give integrated modular relations which are not modular relations in a proper sense, but are their integrated forms. Therefore after differentiation in the parameter, they amount to the ordinary modular relation. However, there is termwise differentiation process involved and we need to make sure that the differentiated series is uniformly convergent as given by [ApoMA, p.403]. The first example is the case where the termwise differentiation is not possible. In [MR, §6.3.3] the theory of arithmetical Fourier series is expounded which may be differentiated termwise with the help of (the equivalent to) the prime number theorem.

In [LWKman] we interpreted the result of Segal [Seg, Theorem 1] as an integrated modular relation.

Theorem 8.1. ([LWKman, Theorem 2]) *For $y > 0$ we have*

$$\int_0^y \widetilde{Q}(x)\,\mathrm{d}x = \sum_{k=1}^{\infty}(1-\cos\frac{y}{k}) \tag{8.1}$$
$$= \frac{\pi}{2}y - \frac{1}{2} + \sqrt{\frac{\pi y}{2}}\sum_{k=1}^{\infty} k^{-1/2} J_1(2\sqrt{2\pi y k}),$$

which is interpreted as $G_{0,1}^{1,0} \leftrightarrow G_{0,2}^{1,0}$.

We recall the process.

The main ingredient is the integrated form of [Seg, (16)]: Let $-1 < b < 0, y > 0$. Then

$$\int_0^y \widetilde{Q}(x)\,\mathrm{d}x = \frac{\pi y}{2\pi i}\int_{(b)} \frac{\Gamma(s)}{\Gamma(2-s)}\zeta(s)(2\pi y)^{-s}\,\mathrm{d}s \tag{8.2}$$

and we transform the right-hand side integral into the series on the right of (8.1).

Applying the asymmetric form of the functional equation

$$\zeta(s) = \pi^{-1}(2\pi)^s \sin\frac{\pi}{2}s\Gamma(1-s)\zeta(1-s), \tag{8.3}$$

we see that the right-hand side of (8.2) becomes

$$-y\frac{1}{2\pi i}\int_{(b)} \Gamma(s-1)\sin\frac{\pi}{2}s\zeta(1-s)y^{-s}\,\mathrm{d}s.$$

Hence substituting the series expression for $\zeta(1-s)$, we conclude that

$$\int_0^y \widetilde{Q}(x)\,\mathrm{d}x = -\sum_{n=1}^{\infty}\frac{y}{n}I_n, \tag{8.4}$$

where

$$\frac{y}{n}I_n = \frac{1}{2\pi i}\int_{(b)} \Gamma(s-1)\sin\frac{\pi}{2}s\left(\frac{y}{n}\right)^{1-s}\mathrm{d}s,$$

say. To express I_n in terms of G-functions, we assure the separation of poles by the path as above (and in [LKT]) and we move the line of integration to $\sigma = c, 1 < c < 2$, thereby encountering a simple pole of the integrand at $s = 1$ with residue 1. Hence it follows that

$$\frac{y}{n} I_n = 1 - \frac{1}{2\pi i} \int_{(c-1)} \Gamma(s) \cos\left(\frac{\pi}{2} s\right) \left(\frac{y}{n}\right)^{-s} ds \tag{8.5}$$
$$= 1 - \left(G_{0,1}^{1,0}\left(e^{\frac{\pi i}{2}} \frac{y}{n} \middle| \begin{matrix} - \\ 0 \end{matrix} \right) + G_{0,1}^{1,0}\left(e^{\frac{-\pi i}{2}} \frac{y}{n} \middle| \begin{matrix} - \\ 0 \end{matrix} \right) \right)$$

since the path $\sigma = c - 1$ separates the poles. The last term reduces to $= 1 - \cos \frac{y}{n}$ in view of

$$G_{1,0}^{0,1}\left(x^{-1} \middle| \begin{matrix} 1 \\ - \end{matrix} \right) = G_{0,1}^{1,0}\left(x \middle| \begin{matrix} - \\ 0 \end{matrix} \right) = \frac{1}{2\pi i} \int_{(\alpha)} x^{-s} \Gamma(s)\, ds = e^{-x} \tag{8.6}$$

valid for $1 > \alpha > 0$ and $\mathrm{Re}\, x \geq 0$ ([ErdH, p. 12, (33)]). Substituting (8.5) in (8.4) proves $G_{0,1}^{1,0}$ part of (8.1).

To prove the $G_{0,2}^{1,0}$ part in (8.1), we shift the line of integration to the right up to $\sigma = c > 1$, thereby encountering simple poles at $s = 0$ and $s = 1$ with residues $\frac{\pi}{2} y$ and $-\frac{1}{2}$, respectively. Hence it follows that

$$\frac{\pi y}{2\pi i} \int_{(b)} \frac{\Gamma(s)}{\Gamma(2-s)} \zeta(s) (2\pi y)^{-s} ds = \frac{\pi}{2} y - \frac{1}{2} + \frac{\pi y}{2\pi i} \int_{(c)} \frac{\Gamma(s)}{\Gamma(2-s)} \zeta(s) (2\pi y)^{-s} ds, \tag{8.7}$$

where we substitute the series expression in the last integral and we are left with the integral

$$\frac{1}{2\pi i} \int_{(c)} \frac{\Gamma(s)}{\Gamma(2-s)} x^{-s} ds = G_{0,2}^{1,0}\left(x \middle| \begin{matrix} - \\ -1, 0 \end{matrix} \right) = x^{-\frac{1}{2}} J_{-1}(2\sqrt{x}).$$

This completes the proof.

8.1. *The* $H_{2,1}^{1,1} \leftrightarrow H_{0,3}^{2,0}$ *formula*

We show that the $H_{2,1}^{1,1} \leftrightarrow H_{0,3}^{2,0}$ formula leads to an identity different from (8.1). Exactly as in Theorem 5.3, we may show that the functional equation

$$\chi(s) = \begin{cases} \Gamma(Cs)\varphi(s), & \mathrm{Re}\, s > \sigma_\varphi \\ \Gamma(C(r-s))\psi(r-s), & \mathrm{Re}\, s < r - \sigma_\psi, \end{cases} \tag{8.8}$$

implies the modular relation

$$\begin{aligned}
&\sum_{k=1}^{\infty} \frac{\alpha_k}{\lambda_k^s} H_{2,1}^{1,1}\left(z\lambda_k \,\middle|\, \begin{matrix}(1-a,A),(b,B)\\ (Cs,C)\end{matrix}\right)\\
&=\sum_{k=1}^{\infty} \frac{\beta_k}{\mu_k^{r-s}} H_{0,3}^{2,0}\left(\frac{\mu_k}{z} \,\middle|\, \begin{matrix}-\\ (C(r-s),C),(a,A),(1-b,B)\end{matrix}\right)\\
&\quad+\sum_{k=1}^{L} \operatorname{Res}\left(\frac{\Gamma(a+As-Aw)}{\Gamma(b-Bs+Bw)}\chi(w)\,z^{s-w}, w=s_k\right).
\end{aligned} \tag{8.9}$$

We apply (8.9) to the following case. Let

$$\varphi(s)=\psi(s)=\sum_{k=1}^{\infty}\frac{\alpha_k}{\lambda_k{}^s} \tag{8.10}$$

and $\alpha_k=\beta_k=\frac{1}{2\sqrt{\pi}}$ and $\lambda_k=\mu_k=\frac{\sqrt{\pi}}{2}k$. Since $\varphi(s)=\frac{2^{s-1}}{\sqrt{\pi}^{s+1}}\zeta(s)$ and $\operatorname{Res}_{s=1}\varphi(s)=\frac{1}{\pi}$. Then the functional equation (8.8) holds with $C=\frac{1}{2}$ and $r=1$. Hence

$$\begin{aligned}
&\frac{1}{4}\left(\frac{2}{\sqrt{\pi}}\right)^{-s-1}\sum_{k=1}^{\infty}\frac{1}{k^s}H_{2,1}^{1,1}\left(z\frac{\sqrt{\pi}}{2}k \,\middle|\, \begin{matrix}(1-a,A),(b,B)\\ (\frac{s}{2},\frac{1}{2})\end{matrix}\right)\\
&\quad+\operatorname{Res}\left(\frac{\Gamma(a+Aw+As-A)}{\Gamma(b-Bw-Bs+B)}\Gamma\left(\tfrac{w}{2}\right)\varphi(w)\,z^{s-1+w}, w=1\right)\\
&=\frac{1}{4}\left(\frac{2}{\sqrt{\pi}}\right)^{2-s}\sum_{k=1}^{\infty}\frac{1}{k^{1-s}}H_{0,3}^{2,0}\left(\frac{\sqrt{\pi}\,k}{2z} \,\middle|\, \begin{matrix}-\\ (\frac{1-s}{2},\frac{1}{2}),(a,A),(1-b,B)\end{matrix}\right)\\
&\quad+\operatorname{Res}\left(\frac{\Gamma(a-Aw+As)}{\Gamma(b+Bw-Bs)}\Gamma\left(\tfrac{w}{2}\right)\varphi(w)\,z^{s-w}, w=1\right).
\end{aligned} \tag{8.11}$$

With $s=1$, $a=A=\frac{1}{2}$, $b=2, B=1$ and z^{-1} for z, (8.11) reads

$$\begin{aligned}
&\frac{\pi^2}{16}\sum_{k=1}^{\infty}\frac{1}{k}H_{2,1}^{1,1}\left(\frac{\sqrt{\pi}}{2z}k \,\middle|\, \begin{matrix}(\frac{1}{2},\frac{1}{2}),(2,1)\\ (\frac{1}{2},\frac{1}{2})\end{matrix}\right)\\
&\quad+\operatorname{Res}\left(\frac{\Gamma(\frac{1}{2}(w+1))}{\Gamma(2-w)}\Gamma\left(\tfrac{w}{2}\right)\varphi(w)\,z^{-w}, w=1\right)\\
&=\frac{1}{2\sqrt{\pi}}\sum_{k=1}^{\infty}H_{0,3}^{2,0}\left(z\frac{\sqrt{\pi}\,k}{2} \,\middle|\, \begin{matrix}-\\ (0,\frac{1}{2}),(\frac{1}{2},\frac{1}{2}),(-1,1)\end{matrix}\right)\\
&\quad+\operatorname{Res}\left(\frac{\Gamma(\frac{1}{2}(2-w))}{\Gamma(1+w)}\Gamma\left(\tfrac{w}{2}\right)\varphi(w)\,z^{w-1}, w=1\right).
\end{aligned} \tag{8.12}$$

Now

$$\frac{1}{2\sqrt{\pi}} H_{0,3}^{2,0}\left(\frac{\sqrt{\pi}\,k}{2} z \,\middle|\, \begin{matrix} - \\ (0,\frac{1}{2}),(\frac{1}{2},\frac{1}{2}),(-1,1) \end{matrix} \right) \tag{8.13}$$
$$= H_{0,2}^{1,0}\left(\sqrt{\pi}\,k\,z \,\middle|\, \begin{matrix} - \\ (0,1),(-1,1) \end{matrix} \right) = G_{0,2}^{1,0}\left(\sqrt{\pi}\,k\,z \,\middle|\, \begin{matrix} - \\ 0,-1 \end{matrix} \right)$$
$$= \frac{1}{\sqrt{\sqrt{\pi}\,k\,z}} J_1\left(2\sqrt{\sqrt{\pi}\,k\,z}\right)$$

by (3.19).

On the other hand,

$$H_{2,1}^{1,1}\left(\frac{\lambda_k}{z} \,\middle|\, \begin{matrix} (\frac{1}{2},\frac{1}{2}),(2,1) \\ (\frac{1}{2},\frac{1}{2}) \end{matrix} \right) = H_{1,2}^{1,1}\left(\frac{z}{\lambda_k} \,\middle|\, \begin{matrix} (\frac{1}{2},\frac{1}{2}) \\ (\frac{1}{2},\frac{1}{2}),(-1,1) \end{matrix} \right) \tag{8.14}$$
$$= \frac{1}{2\pi i}\int_{(c)} \frac{\Gamma\left(\frac{1}{2}+\frac{1}{2}w\right)\Gamma\left(\frac{1}{2}-\frac{1}{2}w\right)}{\Gamma(-1-w)} \left(\frac{z}{\lambda_k}\right)^{-w} dw.$$

We transform the integrand $I(w) := \frac{\Gamma(\frac{1}{2}+\frac{1}{2}w)\Gamma(\frac{1}{2}-\frac{1}{2}w)}{\Gamma(-1-w)}$ in various ways, leading to $\Gamma(2+w)\sin\left(\frac{\pi}{2}w\right)$.

First we multiply by $\frac{\Gamma(2+w)}{\Gamma(2+w)}$, apply the difference equation and the duplication formula to rewrite it as

$$I(w) = \frac{\Gamma\left(\frac{1}{2}+\frac{1}{2}w\right)\Gamma\left(\frac{1}{2}-\frac{1}{2}w\right)\Gamma(2+w)}{\Gamma(-1-w)\Gamma(2+w)} \tag{8.15}$$
$$= -\Gamma(2+w)\frac{\Gamma\left(\frac{1}{2}+\frac{1}{2}w\right)\Gamma\left(\frac{1}{2}-\frac{1}{2}w\right)}{\Gamma(-w)\Gamma(1+w)}$$
$$= -2\pi\Gamma(2+w)\frac{\Gamma\left(\frac{1}{2}+\frac{1}{2}w\right)\Gamma\left(\frac{1}{2}-\frac{1}{2}w\right)}{\Gamma\left(-\frac{1}{2}w\right)\Gamma\left(\frac{1}{2}-\frac{1}{2}w\right)\Gamma\left(\frac{1}{2}+\frac{1}{2}w\right)\Gamma\left(1+\frac{1}{2}w\right)}$$
$$= -2\pi\Gamma(2+w)\frac{1}{\Gamma\left(-\frac{1}{2}w\right)\Gamma\left(1+\frac{1}{2}w\right)},$$

whence it follows that

$$H_{1,2}^{1,1}\left(\frac{z}{\lambda_k} \,\middle|\, \begin{matrix} (\frac{1}{2},\frac{1}{2}) \\ (\frac{1}{2},\frac{1}{2}),(-1,1) \end{matrix} \right) = -2\pi H_{1,2}^{1,0}\left(\frac{z}{\lambda_k} \,\middle|\, \begin{matrix} (1,\frac{1}{2}) \\ (2,1),(1,\frac{1}{2}) \end{matrix} \right) \tag{8.16}$$
$$= \frac{-2\pi}{2\pi i}\left(G_{0,1}^{1,0}\left(e^{\frac{\pi}{2}i}\frac{z}{\lambda_k} \,\middle|\, \begin{matrix} - \\ 2 \end{matrix} \right) - G_{0,1}^{1,0}\left(e^{-\frac{\pi}{2}i}\frac{z}{\lambda_k} \,\middle|\, \begin{matrix} - \\ 2 \end{matrix} \right) \right)$$
$$= i\left(-\frac{z^2}{{\lambda_k}^2} e^{-i\frac{z}{\lambda_k}} + \frac{z^2}{{\lambda_k}^2} e^{-i\frac{z}{\lambda_k}} \right) = 2\frac{z^2}{{\lambda_k}^2}\sin\left(\frac{z}{\lambda_k}\right).$$

Or we apply the reciprocity relation to rewrite it as

$$I(w) = \frac{\Gamma\left(\frac{1}{2}+\frac{1}{2}w\right)\Gamma\left(\frac{1}{2}-\frac{1}{2}w\right)\Gamma(2+w)}{\Gamma(-1-w)\Gamma(2+w)} \tag{8.17}$$

$$= \Gamma(2+w)\frac{\sin(\pi(2+w))}{\pi}\frac{\pi}{\sin\left(\frac{\pi}{2}-\frac{\pi}{2}w\right)}$$

$$= \Gamma(2+w)\sin\left(\frac{\pi}{2}w\right)$$

at which we also arrive from (8.15).

Theorem 8.2.

$$\frac{\pi}{4}z^2\sum_{k=1}^{\infty}\frac{1}{k^3}\sin\left(\frac{2}{\sqrt{\pi}}zk\right) = \frac{1}{2\sqrt{\pi^3}z}\sum_{k=1}^{\infty}\frac{1}{\sqrt{k}}J_1\left(2\sqrt{\sqrt{\pi}\,k\,z}\right)+\mathrm{P}(z), \tag{8.18}$$

where $\mathrm{P}(z) = \frac{1}{\sqrt{\pi}z} - 1$ *is the residual function.*

8.2. *The L-function*

We now consider the Koshlyakov L-function. The relationship between L and K are more intimate than that of X- and Y-functions. The L-function, is so to speak, a co-K function, being related by

$$L_{r_1,r_2}(x) = \frac{1}{2}\left(K_{r_1,r_2}(-ix)+K_{r_1,r_2}(ix)\right) \quad \text{([KoshI, (7.11)])}. \tag{8.19}$$

It is defined by ([KoshI, (8.9)])

$$L(x) = \frac{1}{2\pi i}\int_{(c)}\frac{\pi}{2}G(1-s)\,x^{-s}\mathrm{d}s$$

$$= \frac{1}{2\pi i}\int_{(c)}\frac{\pi}{2}\frac{\Gamma^{r_1}\left(\frac{s}{2}\right)\Gamma^{r_2}(s)}{\Gamma^{r_1}\left(\frac{1-s}{2}\right)\Gamma^{r_2}(1-s)}x^{-s}\mathrm{d}s, \quad c>0,\ x>0. \tag{8.20}$$

The $L_{1,0}(x)$ is similar to $Y_{0,1}(x)$ and

$$L_{1,0}(x) = \frac{\pi}{2}\cdot 2G_{0,2}^{1,0}\left(x^2\,\middle|\,\begin{matrix}-\\0,\frac{1}{2}\end{matrix}\right) = \pi x^{\frac{1}{2}}J_{\frac{1}{2}}(2x)$$

by (3.19), which further reduces to

$$L_{1,0}(x) = \sqrt{\pi}\cos(2x) \tag{8.21}$$

on account of (3.2) (cf. [PBM, p.635,3]).

$L_{0,1}(x)$ is just the same as $Y_{2,0}(x)$ and

$$L_{0,1}(x) = \frac{\pi}{2}G_{0,2}^{1,0}\left(x\,\middle|\,\begin{matrix}-\\0,0\end{matrix}\right) = \frac{\pi}{2}J_0(2\sqrt{x}) \tag{8.22}$$

by (3.19).

Finally, $L_{2,0}(x)$ may be transformed by (1.6) into

$$\begin{aligned} L_{2,0}(x) &= \frac{\pi}{2} H_{0,4}^{2,0}\left(x \,\middle|\, \begin{matrix} - \\ (0,\frac{1}{2}),(0,\frac{1}{2}),(\frac{1}{2},\frac{1}{2}),(\frac{1}{2},\frac{1}{2}) \end{matrix}\right) \\ &= \frac{\pi}{2} 2G_{0,4}^{2,0}\left(x^2 \,\middle|\, \begin{matrix} - \\ 0,0,\frac{1}{2},\frac{1}{2} \end{matrix}\right) \\ &= \pi\left(\frac{2}{\pi}K_0(4\sqrt{x}) - Y_0(4\sqrt{x})\right) \end{aligned}$$

by (9.31), where $Y_s(x)$ signifies the modified Bessel function of the second kind defined by (3.5).

Table 2. Koshlyakov's functions

symb.	rational	imaginary	real
X_{r_1,r_2}	$2e^{-x^2}$	e^{-x}	$4K_0(2x)$
Y_{r_1,r_2}	e^{-x^2}	$\frac{\sin(x)}{x} = \mathrm{si}(x)$	$J_0(2x)$
K_{r_1,r_2}	$\sqrt{\pi}\, e^{-2x}$	$\frac{K_0(2\varepsilon\sqrt{x}) - K_0(2\bar{\varepsilon}\sqrt{x})}{i}$	$2\left(K_0(\varepsilon\sqrt{x}) + K_0(\bar{\varepsilon}\sqrt{x})\right)$
L_{r_1,r_2}	$\sqrt{\pi}\cos(2x)$	$\frac{\pi}{2}J_0(2\sqrt{x})$	$2K_0(4\sqrt{x}) - \pi Y_0(4\sqrt{x})$

In [KoshE], Koshlyakov also mentions:

"It should be noted that deduction of the functional equation for the Dedekind zeta-function from formulas in our general theory is a total vicious circle since the former are already used in deriving the latter."

Such deductions occur

p.129, ll. 1-12 from below

which turns out to be the same as

p.234, ll. 1-10 from below and p.235, ll. 1-16.

However, from our view point, what Koshlyakov did was the deduction of the modular relation from the functional equation and if the reverse deduction is supplied, they will give rise to the equivalent formulations of the functional equation. Cf. e.g. [CJ], [Ham2], [Kob2].

9. Riesz sums

This section is an extract of, and supplement to, Chapter 6 [MR] as well as a supplement to [WW], especially after §9.2 onwards. The Riesz means, or sometimes typical means, were introduced by M. Riesz and have been studied in connection with summability of Fourier series and of Dirichlet series [ChTyp] and [HarRie]. Given an increasing sequence $\{\lambda_k\}$ of real numbers and a sequence $\{\alpha_k\}$ of complex numbers, the Riesz sum of order $\varkappa$ is defined as in [ChTyp, p.2] and [HarRie, p.21] by

$$A^{\varkappa}(x) = A_{\lambda}^{\varkappa}(x) = {\sum_{\lambda_k \le x}}' (x-\lambda_k)^{\varkappa}\alpha_k \tag{9.1}$$
$$= \varkappa \int_0^x (x-t)^{\varkappa-1} A_{\lambda}(t) \mathrm{d}t$$
$$= \varkappa \int_0^x (x-t)^{\varkappa-1} \mathrm{d}A_{\lambda}(t),$$

where

$$A_{\lambda}(x) = A_{\lambda}^{0}(x) = {\sum_{\lambda_k \le x}}' \alpha_k, \tag{9.2}$$

where the prime on the summation sign means that when $\lambda_k = x$, the corresponding term is to be halved.

(9.1) or rather normalized $\frac{1}{\Gamma(\varkappa+1)}A^{\varkappa}(x)$ which appears in (G-8-2) is called the *Riesz sum* of order $\varkappa$ If $\frac{1}{\Gamma(\varkappa+1)}A^{\varkappa}(x)$ approaches a limit A as $x \to \infty$, the sequence $\{\alpha_k\}$ is called Riesz summable or $(R, \varkappa, \lambda)$ summable to A, which is called the *Riesz mean* of the sequence. Sometimes the negative order Riesz sum is considered, in which case the sum is taken over all n which are not equal to x.

In number-theoretic context, it is the Riesz sum rather than the Riesz mean that has been extensively studied. The Riesz sums appear as long as there appears the $G_{1,1}^{1,0}$. Cf. Remark 9.1. There is some mention on the divisor problem in [Ch] in the light of the Riesz sum and there are enormous amount of literature on the Riesz sums and we shall not dwell on well-known cases and state only unexpected use of them.

The application of the Riesz sum comes into play through the Perron formula (9.7) below, sometimes in truncated form. The application of the truncated first order Riesz sum appears on [Dav, p.105] and a truncated general order Riesz sum is treated in [KaR] in both of which the functional equation is not assumed. Riesz sums with the functional equation can be found e.g. in [ChN3], where by differencing, the asymptotic formula for the original sum is deduced. The principle goes back to Landau [LanII] in which one can find the integral order Riesz sum and its reduction to the original partial sum by differencing.

In our unprocessed modular relations, we always have the gamma factor which satisfies the Stirling type growth condition (7.3) and so there is no problem of (absolute) convergence. In the case of the Riesz sum, we must pay attention to the order of the zeta-function such as (7.6) or (9.49). This is why all the authors take higher order Riesz sums and then reduce it the case of the summatory function by differencing (or by differentiating). Stating the results for the summatory function or the 0th order Riesz sum is too hazardous although it sometimes has heuristic meaning. Guessing the expected form and then by integrating it to get into the absolute convergent world and then go down to the desired case is a royal road. Toward the end we shall streamline two cases of such unsound applications of the Riesz sums. Unlike the case of arithmetical Fourier series in [MR, §6.3.3], where in some cases one needs to make a recourse to the prime number theorem to assure the uniform convergence, the Riesz sum case is sometimes simpler and the above reduction often succeeds.

The general formula for the difference operator of order $\alpha \in \mathbb{N}$ with difference $y \geq 0$ is given by

$$\Delta_y^\alpha f(x) = \sum_{\nu=0}^{\alpha} (-1)^{\alpha-\nu} \binom{\alpha}{\nu} f(x+\nu y). \tag{9.3}$$

If f has the α-th derivative $f^{(\alpha)}$, then

$$\Delta_y^\alpha f(x) = \int_x^{x+y} \mathrm{d}t_1 \int_{t_1}^{t_1+y} \mathrm{d}t_2 \cdots \int_{t_{\alpha-1}}^{t_{\alpha-1}+y} f^{(\alpha)}(t_\alpha)\, \mathrm{d}t_\alpha. \tag{9.4}$$

The Riesz kernel which produces the Riesz sum is defined by

$$G_{1,1}^{1,0}\left(z \,\middle|\, \begin{matrix} a \\ b \end{matrix}\right) = \begin{cases} \dfrac{1}{\Gamma(a-b)} z^b (1-z)^{a-b-1}, & |z|<1 \\ \dfrac{1}{2} & z=1, a=b+1 \\ 0, & |z|>1. \end{cases} \tag{9.5}$$

Remark 9.1. Notes on (9.5). Let $\varkappa \geq 0$ denote the order of the Riesz mean and set $b=0, a=\varkappa+1$. Then (9.5) reads ($c>0$)

$$\frac{1}{2\pi i}\int_c \frac{\Gamma(s)}{\Gamma(s+\varkappa+1)} z^{-s}\, \mathrm{d}s = \begin{cases} \dfrac{1}{\Gamma(\varkappa+1)}(1-z)^\varkappa, & (|z|<1) \\ \frac{1}{2}, & (\varkappa=0, z=1) \\ 0, & (|z|>1). \end{cases} \tag{9.6}$$

This can be found in Hardy-Riesz [HarRie] and Chandrasekharan and Minakshisundaram [ChTyp] and used in the context of *Perron's formula*

$$\frac{1}{\Gamma(\varkappa+1)}\sum_{\lambda_k\le x}{}' \alpha_k(x-\lambda_n)^{\varkappa} = \frac{1}{2\pi i}\int_c \frac{\Gamma(s)\varphi(s)x^{s+\varkappa}}{\Gamma(s+\varkappa+1)}\,\mathrm{d}s, \tag{9.7}$$

where the left-hand side sum is called the *Riesz sum* of order $\varkappa$ and denoted $A_\lambda^\varkappa(x)$ as mentioned above and $\varphi(s) = \sum_{k=1}^{\infty}\frac{\alpha_k}{k^s}$. We note that the exponent of x is s rather than $-s$, so that in the G-function setting, x always appears as its reciprocal.

If the order $\varkappa \in \mathbb{N}$, then the right-hand side member of (9.6) is

$$\frac{1}{2\pi i}\int_c \frac{1}{s(s+1)\cdots(s+\varkappa)}z^{-s}\,\mathrm{d}s$$

and the Riesz sum amounts to the $\varkappa$ times integration of the original sum $A_\lambda(x)$. Thus Landau's differencing is an analogue of the integration and differentiation.

In view of this integration-differentiation aspect there are a number of cases in which the Riesz sum appears in disguised form. Especially, when there is a gamma factor $\frac{\Gamma(s)}{\Gamma(s+\varkappa+1)}$ or $\binom{\varkappa+1}{0}$ involved.

The special case of (9.6) with $\varkappa = 0$ is known as the *discontinuous integral* whose truncated form can be found e.g. in Davenport [Dav, pp. 109-110].

The very special case $z = 1, a-b = 1$ of (9.5) (and of the corresponding logarithmic case $(z = 1, m = 1)$) presents excessive complexities in notation, so that we follow Hardy and Riesz [HarRie] to use (9.7) by suppressing the prime on the summation. Below we sometimes add the prime to emphasize the summatory function.

9.1. *Improper modular relations as Riesz sums*

In [LWKman] we introduced the notion of an *improper modular relation* to the effect that the identity in question does not look like a modular relation because some terms are placed on the other side as part of the residues, say or for some other reasons.

Theorem 9.1. ([Kat2, Theorem 4.1]) *For all $x \ge |\mathrm{Re}\,\nu| + 2$ and $\nu \notin \mathbb{Z}$,*

$$\sum_{n=0}^{\infty}(-1)^n\binom{x}{n}\zeta(n-\nu) = \frac{\Gamma(x+1)\,\Gamma(-\nu-1)}{\Gamma(x-\nu)} + \mathcal{H}_\nu(x), \tag{9.8}$$

where

$$\mathcal{H}_\nu(x) = \Gamma(x+1)\sum_{n=1}^{\infty}\Big(U(x+1,\nu+2;-2\pi in)+U(x+1,\nu+2;2\pi in)\Big). \quad (9.9)$$

In the case $\nu \in \mathbb{Z}$, $\nu \geq -1$, *the first term on the right of* (9.8) *is to be replaced by*

$$(-1)^\nu\binom{x}{\nu+1}\left(\frac{\Gamma'}{\Gamma}(x-\nu)+2\gamma-H_{\nu+1}\right), \quad (9.10)$$

where γ *is Euler's constant and* $H_{\nu+1}$ *is the* $(\nu+1)$*-th harmonic number* $H_{\nu+1} = \sum_{n=1}^{\nu+1}\frac{1}{n}$.

Remark 9.2. In the case $\operatorname{Re}\nu+1 \geq 0$, the parameter ν affects the series and we are to shift the first $[\operatorname{Re}\nu+1]$ partial sum to the right so as to indicate that these $\zeta(n-\nu)$'s are not the series but its analytic continuation, so that the right-hand side is to mean

$$H_\nu(x) = \frac{\Gamma(x+1)\,\Gamma(-\nu-1)}{\Gamma(x-\nu)} - \sum_{n=0}^{[\operatorname{Re}\nu+1]}(-1)^n\binom{x}{n}\zeta(n-\nu)+\mathcal{H}_\nu(x) \quad (9.11)$$

for $\nu \notin \mathbb{Z}$; for $\nu \in \mathbb{Z}$ we are to replace the first term in (9.11) by (9.10). On the other hand, the left-hand side sum starts from $n > [\operatorname{Re}\nu+1]$ and all the work on these problems have this type of sums.

We refer to this as the improper modular relations. They do not look like at a first glance a modular relation but even in the case $\operatorname{Re}\nu+1 > 0$, when stated in the form of our Theorem 9.1, it is a modular relation as furnished by Theorem 9.2 below.

In much the same way, we may interpret other improper modular relations ([Kat1, Theorems 3.1, 3.2]) of Katsurada as modular relations (cf. Theorem 9.2). The following is due to Wang and Wang [WW].

Example 9.1 (Katsurada's formula [Kat1]). *Setting* $a = 0, A = 1, b = \frac{s}{2}, B = \frac{1}{2}$ *in* (8.9), *we deduce the first generalized Katsurada formula*

$$\begin{aligned}
&\sum_{k=1}^{\infty} \frac{\alpha_k}{\lambda_k^s} \exp\left(-\frac{1}{z\lambda_k}\right) \\
&= \frac{2}{\sqrt{\pi}} \sum_{k=1}^{\infty} \beta_k \left\{ \left(\frac{e^{\frac{\pi}{2}i}}{2z\mu_k}\right)^{\frac{1-s}{2}} K_{1-s}\left(2\sqrt{2}\, e^{-\frac{\pi}{4}i} \sqrt{\frac{\mu_k}{z}}\right) \right. \\
&\qquad \left. + \left(\frac{e^{-\frac{\pi}{2}i}}{2z\mu_k}\right)^{\frac{1-s}{2}} K_{1-s}\left(2\sqrt{2}\, e^{\frac{\pi}{4}i} \sqrt{\frac{\mu_k}{z}}\right) \right\} \\
&\quad + \sum_{k=1}^{L} \operatorname{Res}\left(\frac{\Gamma(s-w)}{\Gamma\left(\frac{w}{2}\right)} \chi(w)\, z^{s-w}, w = s_k\right),
\end{aligned} \tag{9.12}$$

which in the case of the Riemann zeta-function, reduces to (9.16).

Theorem 9.2. *The Riesz sum formula, a special case of the* $H_{2,2}^{1,1} \leftrightarrow H_{1,3}^{2,0}$ *formula,*

$$\begin{aligned}
&\frac{1}{\pi^{\frac{s}{2}}} \sum_{k=1}^{\infty} \frac{1}{k^s} G_{1,1}^{1,0}\left(\frac{1}{z} \,\middle|\, \begin{matrix} 1-b \\ 0 \end{matrix}\right) \\
&\quad + \operatorname{Res}\left(\frac{\Gamma(w+s-1)}{\Gamma\left(-\frac{w}{2}+\frac{1}{2}\right)\Gamma(1-b+w+s-1)} \right. \\
&\qquad\qquad \left. \Gamma\left(\frac{w}{2}\right)\varphi(w)\, z^{s-1+w}, w = 1\right) \\
&= \frac{1}{\pi^{\frac{1-s}{2}}} \sum_{k=1}^{\infty} \frac{1}{k^{1-s}} H_{1,3}^{2,0}\left(\frac{\sqrt{\pi}\,k}{z} \,\middle|\, \begin{matrix} (1-b,1) \\ \left(\frac{1-s}{2},\frac{1}{2}\right),(0,1),\left(1-\frac{s}{2},\frac{1}{2}\right) \end{matrix}\right) \\
&\quad + \operatorname{Res}\left(\frac{\Gamma(-w+s)}{\Gamma\left(\frac{w}{2}\right)\Gamma(1-b-w+s)} \Gamma\left(\frac{w}{2}\right)\varphi(w)\, z^{s-w}, w = 1\right)
\end{aligned} \tag{9.13}$$

in the limiting case as $1-b \to \infty$ *amounts to Katsurada's first formula ([Kat1, Theorems 3.1, 3.2])*

$$\begin{aligned}
&\sum_{n=0}^{\infty} (-1)^n \frac{x^n}{n!} \zeta(n-\nu) = \sum_{k=1}^{\infty} k^{\nu} e^{-\frac{x}{k}} \\
&= 2\sum_{k=1}^{\infty} \left\{ \left(\frac{x}{2\pi i k}\right)^{\frac{\nu+1}{2}} K_{\nu+1}\left(2\sqrt{2\pi i k x}\right) \right. \\
&\qquad \left. + \left(-\frac{x}{2\pi i k}\right)^{\frac{\nu+1}{2}} K_{\nu+1}\left(2\sqrt{-2\pi i k x}\right) \right\} + x^{\nu+1}\,\Gamma(-\nu-1)
\end{aligned} \tag{9.14}$$

and in the case $z \to 1$ to Katsurada's second formula ([Kat1, Theorem 4.1])

$$\begin{aligned}
&\frac{1}{\Gamma(1+x)} \sum_{n=0}^{\infty} \frac{(x)_n}{n!} \zeta(n-\nu) \\
&= \sum_{k=1}^{\infty} \Big\{ U\big(x+1, \nu+2; 2\pi i k\big) + U\big(x+1, \nu+2; -2\pi i k\big) \Big\} \\
&\quad + \frac{\Gamma(-\nu-1)}{\Gamma(x-\nu)}.
\end{aligned} \tag{9.15}$$

Proof. The first equality in (9.14) follows by expanding into the Taylor series, changing the order of summation and writing the inner sum as $\zeta(n-\nu)$.

Hence in order to prove (9.14), it suffices to prove that

$$\begin{aligned}
&\sum_{k=1}^{\infty} k^{\nu} e^{-\frac{x}{k}} \\
&= 2\sum_{k=1}^{\infty} \Big\{ \Big(\frac{x}{2\pi i k}\Big)^{\frac{\nu+1}{2}} K_{\nu+1}\Big(2\sqrt{2\pi i k x}\Big) \\
&\quad + \Big(-\frac{x}{2\pi i k}\Big)^{\frac{\nu+1}{2}} K_{\nu+1}\Big(2\sqrt{-2\pi i k x}\Big) \Big\} + x^{\nu+1}\,\Gamma(-\nu-1).
\end{aligned} \tag{9.16}$$

The special case of (9.24) with $C = \frac{1}{2}$, $a_1 = 0$, $a_2 = \frac{s}{2}$, $b_1 = 1-b$, $A_1 = 1$, $A_2 = \frac{1}{2}$, and $B_1 = 1$ leads to (9.13) in view of the reduction formula. Although $G^{1,0}_{1,1}\left(\frac{1}{z} \,\middle|\, \begin{matrix} 1-b \\ 0 \end{matrix}\right)$ is the Riesz kernel (9.5) and in most cases one considers the Riesz sum for $k \le x$ by taking $z = \frac{x}{k}$ and $x \ge 1$, the exceptional case $0 < x < 1$ is also considered pertaining to (9.15), in which case the left-hand side reads $\frac{1}{\Gamma(1-b)} \sum_{k=1}^{\infty} k^{\nu}\left(1-\frac{x}{k}\right)^{-b}$. Hence (9.13) amounts in the case $|z| < 1$ to

$$\begin{aligned}
&\frac{1}{\Gamma(1-b)} \sum_{k=1}^{\infty} k^{\nu}\Big(1-\frac{z}{k}\Big)^{-b} \\
&= x^{\nu+1} \sum_{k=1}^{\infty} \Big\{ e^{-2\pi i k x}\, U\big(1-b, \nu+2; 2\pi i k z\big) \\
&\qquad + e^{2\pi i k z}\, U\big(1-b, \nu+2; -2\pi i k x\big) \Big\} + z^{\nu+1} \frac{\Gamma(-\nu-1)}{\Gamma(-\nu-b)}.
\end{aligned} \tag{9.17}$$

Rewriting (9.17) as

$$\sum_{k=1}^{\infty} k^{\nu}\left(1-\frac{\frac{x}{k}}{1-b}\right)^{1-b} \tag{9.18}$$

$$= \frac{\Gamma(1-b)}{\Gamma(-\nu-b)(1-b)^{\nu+1}} x^{\nu+1} \sum_{k=1}^{\infty} \Big\{ e^{-2\pi i k x}\, U\big(1-b, \nu+2; 2\pi i k \frac{x}{1-b}\big)$$

$$+ e^{2\pi i k \frac{x}{1-b}}\, U\big(1-b, \nu+2; -2\pi i k x\big) + \Gamma(-\nu-1)\Big\},$$

we take the limit as $1-b \to \infty$ to conclude (9.14) for $|z|<1$, thereby incorporating the following limit relations.

$$\lim_{a\to\infty}\left(\Gamma(a-c+1)U\left(a,c;\frac{x}{a}\right)\right) = 2x^{\frac{1}{2}(1-c)}K_{c-1}(2\sqrt{x}), \tag{9.19}$$

which is [ErdH, I,(19),p.266];

$$\lim_{1-b\to\infty} \frac{\Gamma(1-b)}{\Gamma(-\nu-b)(1-b)^{\nu+1}} = 1, \tag{9.20}$$

and

$$\lim_{1-b\to\infty}\left(1-\frac{\frac{x}{k}}{1-b}\right)^{1-b} = e^{-\frac{x}{k}}. \tag{9.21}$$

The restriction $|z|<1$ may be removed by analytic continuation and the proof follows for other values of z.

Expanding into the binomial series, changing the order of summation and writing the inner sum as $\zeta(n-\nu)$ (note the similarity of the procedure as in the proof of the first equality in (9.14)), we immediately see that the left-hand side of (9.17) is

$$\frac{1}{\Gamma(1+x)} \sum_{n=0}^{\infty} \frac{(x)_n}{n!}\, \zeta(n-\nu)\, z^n,$$

so that (9.17) reads

$$\frac{1}{\Gamma(1+x)} \sum_{n=0}^{\infty} \frac{(x)_n}{n!}\, \zeta(n-\nu)\, z^n \tag{9.22}$$

$$= x^{\nu+1} \sum_{k=1}^{\infty} \Big\{ e^{-2\pi i k x}\, U\big(1-b, \nu+2; 2\pi i k z\big)$$

$$+ e^{2\pi i k z}\, U\big(1-b, \nu+2; -2\pi i k x\big)\Big\} + z^{\nu+1} \frac{\Gamma(-\nu-1)}{\Gamma(-\nu-b)},$$

whence (9.15) immediately follows. □

Remark 9.3. Another way would be to study the Riesz sum or the Riesz mean and it is well-known that the Riesz summability implies Abel summability [HarRie, Theorem 24]. There is an explicit formula known for the transition.

Lemma 9.1. *The sum of the Abel mean* $\sum_{k=1}^{\infty} a_n e^{-\lambda_k s}$ *at all points of the sector* $|\arg s| \le u < \frac{\pi}{2}$ *other than the origin is*

$$\frac{1}{\Gamma(\varkappa+1)} \int_0^{\infty} s^{\varkappa+1} e^{-s\tau} A^{\varkappa}(\tau)\,\mathrm{d}\tau, \tag{9.23}$$

where $A^{\varkappa}(x)$ *is the* $\varkappa$*-th Riesz sum defined in* (9.1).

9.2. The $H_{2,2}^{1,1} \leftrightarrow H_{1,3}^{2,0}$ formula

As a special case of Theorem 5.3 with Hecke type functions equation, we have

$$\begin{aligned}
&\sum_{k=1}^{\infty} \frac{\alpha_k}{\lambda_k^s} H_{2,2}^{1,1}\left(z\lambda_k \,\middle|\, \begin{matrix} (1-a_1, A_1), (a_2, A_2) \\ (Cs, C), (1-b_1, B_1) \end{matrix} \right) \\
&= \sum_{k=1}^{\infty} \frac{\beta_k}{\mu_k^{r-s}} H_{1,3}^{2,0}\left(\frac{\mu_k}{z} \,\middle|\, \begin{matrix} (b_1, B_1) \\ (C(r-s), C), (a_1, A_1), (1-a_2, A_2) \end{matrix} \right) \\
&\quad + \sum_{k=1}^{L} \mathrm{Res}\left(\frac{\Gamma(a_1 + A_1 s - A_1 w)}{\Gamma(a_2 - A_2 s + A_2 w)\,\Gamma(b_1 + B_1 s - B_1 w)} \chi(w)\, z^{s-w}, w = s_k \right).
\end{aligned} \tag{9.24}$$

Example 9.2 (Katsurada's formula [Kat1] again). *Consider the case of the Riemann zeta-function* $C = \frac{1}{2}$. *By setting* $a_1 = 0$, $a_2 = \frac{s}{2}$, $b_1 = 1-b$, $A_1 = 1$, $A_2 = \frac{1}{2}$ *and* $B_1 = 1$ *in* (9.24), *deduce the second generalized Katsurada formula*

$$\begin{aligned}
&\frac{1}{\pi^{\frac{s}{2}}} \sum_{k=1}^{\infty} \frac{1}{k^s} H_{2,2}^{1,1}\left(z\sqrt{\pi}\,k \,\middle|\, \begin{matrix} (1,1), (\frac{s}{2}, \frac{1}{2}) \\ (\frac{s}{2}, \frac{1}{2}), (b, 1) \end{matrix} \right) \\
&\quad + \mathrm{Res}\Bigg(\frac{\Gamma(w+s-1)}{\Gamma(-\frac{w}{2}+\frac{1}{2})\,\Gamma(1-b+w+s-1)} \\
&\qquad\qquad \Gamma(\tfrac{w}{2}) Z(w)\, z^{s-1+w}, w = 1 \Bigg) \qquad (9.25) \\
&= \frac{1}{\pi^{\frac{1-s}{2}}} \sum_{k=1}^{\infty} \frac{1}{k^{1-s}} H_{1,3}^{2,0}\left(\frac{\sqrt{\pi}\,k}{z} \,\middle|\, \begin{matrix} (1-b, 1) \\ (\frac{1-s}{2}, \frac{1}{2}), (0,1), (1-\frac{s}{2}, \frac{1}{2}) \end{matrix} \right) \\
&\quad + \mathrm{Res}\left(\frac{\Gamma(-w+s)}{\Gamma(\frac{w}{2})\,\Gamma(1-b-w+s)} \Gamma(\tfrac{w}{2}) Z(w)\, z^{s-w}, w = 1 \right),
\end{aligned}$$

which in particular entails another form of (9.17)

$$
\begin{aligned}
&\frac{1}{\Gamma(1-b)}\sum_{n=0}^{\infty}\frac{(b)_n}{n!}\,\zeta(n-\nu)\,x^n \\
&= x^{\nu+1}\sum_{k=1}^{\infty}\Big\{e^{-2\pi ikx}\,\Psi\big(1-b,\nu+2;2\pi ikx\big) \\
&\qquad + e^{2\pi ikx}\,\Psi\big(1-b,\nu+2;-2\pi ikx\big)\Big\} + x^{\nu+1}\,\frac{\Gamma(-\nu-1)}{\Gamma(-\nu-b)}.
\end{aligned}
\tag{9.26}
$$

Proof On the left-hand side, $H_{1,2}^{1,1}$ reduces to the Riesz kernel while on the right-hand side

$$
\begin{aligned}
&H_{1,3}^{2,0}\left(z\,\middle|\,\begin{matrix}(1-b,1)\\ \big(s,\tfrac12\big),(0,1),\big(s+\tfrac12,\tfrac12\big)\end{matrix}\right) \\
&= \frac{z^{2s}}{\sqrt{\pi}}\Big\{e^{2iz}\,U\big(1-b,2s+1;-2iz\big) + e^{-2iz}\,U\big(1-b,2s+1;2iz\big)\Big\}
\end{aligned}
\tag{9.27}
$$

after some transformations.

9.3. *Generalization of Wilton's Riesz sum formula*

The most far-reaching Riesz sum result was obtained by J. R. Wilton [Wil2]. To describe it we need some auxiliary results which are of interest in their own right and may find other places of application. For details we refer to [MR, §9.1.2].

Theorem 9.3. *We have*

$$
\begin{aligned}
&G_{1,3}^{2,0}\left(z\,\middle|\,\begin{matrix}c\\ a,b,c\end{matrix}\right) \\
&= z^{\frac12(a+b)}\Big\{-\sin((c-b)\pi)\,Y_{a-b}\big(2\sqrt{z}\,\big) + \cos((c-b)\pi)\,J_{a-b}\big(2\sqrt{z}\,\big)\Big\}
\end{aligned}
\tag{9.28}
$$

and

$$G_{2,6}^{4,0}\left(z\,\middle|\,\begin{matrix} c,d \\ a,b,a+\frac{1}{2},b+\frac{1}{2},c,d\end{matrix}\right)$$
$$=\frac{1}{4^{a+b}}\left\{\frac{\cos((c-d)\pi)}{\pi}\,G_{0,2}^{2,0}\left(4\sqrt{z}\,\middle|\,\begin{matrix} - \\ 2a,2b\end{matrix}\right)\right.$$
$$\left.+\,G_{1,3}^{2,0}\left(4\sqrt{z}\,\middle|\,\begin{matrix} c+d-\frac{1}{2} \\ 2a,2b,c+d-\frac{1}{2}\end{matrix}\right)\right\}$$
$$=z^{\frac{1}{2}(a+b)}\left\{\frac{2}{\pi}\cos((c-d)\pi)\,K_{2a-2b}\big(4\sqrt[4]{z}\,\big)\right.$$
$$\left.+\cos((c+d-2b)\pi)\,Y_{2a-2b}\big(4\sqrt[4]{z}\,\big)+\sin((c+d-2b)\pi)\,J_{2a-2b}\big(4\sqrt[4]{z}\,\big)\right\}. \tag{9.29}$$

Proof is omitted.

Corollary 9.1.

(1) By setting $c=a+\frac{1}{2}$, $d=b+\frac{1}{2}$, *we have*

$$G_{0,4}^{2,0}\left(z\,\middle|\,\begin{matrix} - \\ a,b,a+\frac{1}{2},b+\frac{1}{2}\end{matrix}\right) \tag{9.30}$$
$$=z^{\frac{1}{2}(a+b)}\left\{\frac{2}{\pi}\cos((a-b)\pi)\,K_{2a-2b}\big(4\sqrt[4]{z}\,\big)-\cos((a-b)\pi)\,Y_{2a-2b}\big(4\sqrt[4]{z}\,\big)\right.$$
$$\left.-\sin((a-b)\pi)\,J_{2a-2b}\big(4\sqrt[4]{z}\,\big)\right\}$$

and in particular

$$G_{0,4}^{2,0}\left(z\,\middle|\,\begin{matrix} - \\ a,a,a+\frac{1}{2},a+\frac{1}{2}\end{matrix}\right)=z^{a}\left\{\frac{2}{\pi}K_0\big(4\sqrt[4]{z}\,\big)-Y_0\big(4\sqrt[4]{z}\,\big)\right\}. \tag{9.31}$$

(2) By setting $c=a+\frac{1}{2}$, $d=b$, *we have*

$$G_{0,4}^{2,0}\left(z\,\middle|\,\begin{matrix} - \\ a,b+\frac{1}{2},a+\frac{1}{2},b\end{matrix}\right) \tag{9.32}$$
$$=z^{\frac{1}{2}(a+b)}\left\{-\frac{2}{\pi}\sin((a-b)\pi)\,K_{2a-2b}\big(4\sqrt[4]{z}\,\big)-\sin((a-b)\pi)\,Y_{2a-2b}\big(4\sqrt[4]{z}\,\big)\right.$$
$$\left.+\cos((a-b)\pi)\,J_{2a-2b}\big(4\sqrt[4]{z}\,\big)\right\}.$$

(3) By setting $c = a$, $d = b$, *we have*

$$G^{2,0}_{0,4}\left(z \,\middle|\, \begin{matrix} - \\ a+\frac{1}{2}, b+\frac{1}{2}, a, b \end{matrix}\right) \tag{9.33}$$
$$= z^{\frac{1}{2}(a+b)} \left\{ \frac{2}{\pi} \cos((a-b)\pi)\, K_{2a-2b}\big(4\sqrt[4]{z}\,\big) + \cos((a-b)\pi)\, Y_{2a-2b}\big(4\sqrt[4]{z}\,\big) + \sin((a-b)\pi)\, J_{2a-2b}\big(4\sqrt[4]{z}\,\big) \right\}$$

and in particular

$$G^{2,0}_{0,4}\left(z \,\middle|\, \begin{matrix} - \\ a+\frac{1}{2}, a+\frac{1}{2}, a, a \end{matrix}\right) = z^a \left\{ \frac{2}{\pi} K_0\big(4\sqrt[4]{z}\,\big) + Y_0\big(4\sqrt[4]{z}\,\big) \right\}. \tag{9.34}$$

Slightly more general than Wilton's [Wil4, (1.22)], we introduce *Wilton's generalized Bessel function*

$$\begin{aligned} G^\lambda_\nu(z) = &-(-1)^\lambda \frac{2}{\pi} \sin\left(\frac{\nu-\lambda}{2}\pi\right) K_\nu(z) \\ &- \sin\left(\frac{\nu-\lambda}{2}\pi\right) Y_\nu(z) + \cos\left(\frac{\nu-\lambda}{2}\pi\right) J_\nu(z). \end{aligned} \tag{9.35}$$

Then the following generalization of Wilton's theorem [Wil4, Theorem 1] holds true.

Theorem 9.4. (Generalization of Wilton's Riesz sum) *For a non-negative integer* $\varkappa$ *and* $z > 0$, *we have the Riesz sum formula*

$$\frac{2^{\varkappa+1} z^{-\nu-\varkappa}}{\Gamma(\varkappa+1)} \sum_{\lambda_k < z} \alpha_k \lambda_k^\nu (z-\lambda_k)^\varkappa \tag{9.36}$$
$$= 2\, z^{-\frac{\lambda-1}{2}} \sum_{k=1}^\infty \frac{\beta_k}{\mu_k^{\frac{\varkappa+1}{2}}} G^\lambda_{2\nu+\lambda+1}\left(4\sqrt{z\mu_k}\right)$$
$$+ \sum_{k=1}^L \operatorname{Res}\left(\frac{\Gamma\big(-\frac{\nu}{2}-\frac{\varkappa}{2}+\frac{1}{2}-\frac{w}{2}\big)\,\Gamma\big(-\frac{\nu}{2}-\frac{\varkappa}{2}-\frac{w}{2}\big)}{\Gamma\big(-\frac{\nu}{2}+\frac{w}{2}\big)\,\Gamma\big(\frac{\nu}{2}+\varkappa+1+\frac{w}{2}\big)\,\Gamma\big(-\frac{\nu}{2}+\frac{1}{2}-\frac{w}{2}\big)\,\Gamma\big(-\frac{\nu}{2}-\varkappa-\frac{w}{2}\big)} \chi(w)\, z^w, w = s_k \right),$$

where $G^\varkappa_\nu(z)$ *is Wilton's generalized Bessel function* (9.35).

The special case of $\varkappa = 0$ of (9.36) reads

Corollary 9.2.

$$
\begin{aligned}
&2\, z^{-\nu} \sideset{}{'}\sum_{\lambda_k \le z} \alpha_k \lambda_k^\nu \\
&= 2\, z^{\frac{1}{2}} \sum_{k=1}^{\infty} \frac{\beta_k}{\sqrt{\mu_k}} F_{2\nu+1}\Big(4\sqrt{z\mu_k}\Big) \qquad (9.37)\\
&\quad + \sum_{k=1}^{L} \mathrm{Res}\left(\frac{1}{\Gamma\big(\frac{\nu}{2} + \frac{w}{2} + 1\big)\,\Gamma\big(-\frac{\nu}{2} + \frac{w}{2}\big)}\, \chi(w)\, z^w, w = s_k \right),
\end{aligned}
$$

where

$$
F_\nu(z) = G_\nu^0(z) = -\frac{2}{\pi} \sin\Big(\frac{\nu}{2}\pi\Big) K_\nu(z) - \sin\Big(\frac{\nu}{2}\pi\Big) Y_\nu(z) + \cos\Big(\frac{\nu}{2}\pi\Big) J_\nu(z). \qquad (9.38)
$$

By Convention, Theorem 9.4 reduces to the *Oppenheim-Wilton's formula* (for $x > 0$)

$$
\begin{aligned}
&\frac{2^{\varkappa+1}}{\Gamma(\varkappa+1)} \sideset{}{'}\sum_{k \le x} \sigma_{2\nu}(k)(x-k)^{\varkappa} \qquad (9.39)\\
&= 2\pi^{-\varkappa}\, x^{\nu+\frac{\varkappa}{2}+\frac{1}{2}} \sum_{k=1}^{\infty} \frac{\sigma_{2\nu}(k)}{k^{\nu+\frac{\varkappa}{2}+\frac{1}{2}}} G_{2\nu+\varkappa+1}^{\lambda}\Big(4\pi\sqrt{kz}\Big)\\
&\quad + \sum_{v \in S_\nu} \mathrm{Res}\left(\frac{\Gamma\big(-\frac{\nu}{2} - \frac{\varkappa}{2} + \frac{1}{2} - \frac{w}{2}\big)\,\Gamma\big(-\frac{\nu}{2} - \frac{\varkappa}{2} - \frac{w}{2}\big)}{\Gamma\big(\frac{\nu}{2} + \varkappa + 1 + \frac{w}{2}\big)\,\Gamma\big(-\frac{\nu}{2} + \frac{1}{2} - \frac{w}{2}\big)\,\Gamma\big(-\frac{\nu}{2} - \varkappa - \frac{w}{2}\big)} \right.\\
&\qquad\qquad \left. \Gamma\big(\tfrac{w+\nu}{2}\big)\, \zeta(w+\nu)\, \zeta(w-\nu)\, x^{\nu+\varkappa+w}, w = v \right),
\end{aligned}
$$

while Corollary 9.2 reduces to

$$
\begin{aligned}
\sideset{}{'}\sum_{k \le x} \sigma_{2\nu}(k) &= x^{\frac{2\nu+1}{2}} \sum_{k=1}^{\infty} \frac{\sigma_{2\nu}(k)}{k^{\nu+\frac{1}{2}}} F_{2\nu+1}\Big(4\pi\sqrt{xk}\Big)\\
&\quad + x^\nu \sum_{v \in S_\nu} \mathrm{Res}\left(\frac{\zeta(w+\nu)\,\zeta(w-\nu)}{w+\nu}\, x^w, w = v \right).
\end{aligned} \qquad (9.40)
$$

We may easily calculate the residues and we have

$$
\mathrm{P}_\nu(x) = \zeta(1-2\nu)x + \frac{\zeta(2\nu+1)}{2v+1} x^{2\nu+1} - \frac{1}{2}\zeta(-2\nu) \qquad (9.41)
$$

for $\nu \ne 0, \pm\frac{1}{2}$,

$$
\mathrm{P}_0(x) = x\log x + (2\gamma - 1)x + \frac{1}{4} \qquad (9.42)
$$

and

$$\mathrm{P}_{1/2}(x) = -\frac{1}{12}(\sqrt{x} + \frac{1}{\sqrt{x}}), \mathrm{P}_{1/2}(x) = \frac{\pi^2}{6}x^{\frac{3}{2}} - \frac{1}{12}\sqrt{z}(\log x + \gamma). \quad (9.43)$$

(9.40) together with (9.41) coincide with [AAW4, pp.9-10] and especially, the function $F(w)$ on [AAW4, ll.1-2,p.10] is $w^{\frac{\nu}{2}}F_\nu(4\sqrt{w})$.

(9.40) with $\nu = 0$ together with (9.42) coincide with [AAW1, (16)] and especially, the function $F(w)$ on [AAW1, (17)] is $w^{\frac{1}{2}}F_\nu(4\sqrt{w})$. This is the celebrated *Voronoĭ formula* ([Vor1])

$$\begin{aligned}\sum_{k\le x}{}' d(k) &= x\log x + (2\gamma - 1)x + \frac{1}{4} \\ &- \sqrt{x}\sum_{k=1}^{\infty}\frac{d(k)}{\sqrt{k}}\left(Y_1\left(4\pi\sqrt{xk}\right) + \frac{2}{\pi}K_1\left(4\pi\sqrt{xk}\right)\right).\end{aligned} \quad (9.44)$$

In exactly the same way, we obtain the formula for the summatory function of the coefficients of the Dedekind zeta-function of a real quadratic field [AAW1, (18)]

$$\begin{aligned}&\sum_{k\le x}{}' F(k) \\ &= \rho x - \sqrt{x}\sum_{k=1}^{\infty}\frac{F(k)}{\sqrt{k}}\left(Y_1\left(4\pi\sqrt{\frac{xk}{|\Delta|}}\right) + \frac{2}{\pi}K_1\left(4\pi\sqrt{\frac{xk}{|\Delta|}}\right)\right),\end{aligned} \quad (9.45)$$

where the residue is defined by (2.4).

9.4. *Linearized product of two zeta-functions*

In this subsection we shall elucidate the results of Nakajima [Nak2] which are generalizations of identities obtained by [Wil2] and further developed by Bellman [Bell]. The results are due to [DeM] and also incorporated in [MR, Chapter 6] and we shall be brief. The results are obtained on the basis of the Atkinson dissection [Atk2], where the *Atkinson dissection* indicates the splitting of the double sum:

$$\sum_{m,n=1}^{\infty} = \sum_{m=1}^{\infty}\sum_{n<m} + \sum_{m=n=1}^{\infty} + \sum_{n=1}^{\infty}\sum_{m<n}, \quad (9.46)$$

which is originally due to Euler. Nakajima views this as the sum of series involving discontinuous integrals rather than the Perron formula as he claims.

We consider the Dirichlet series $\varphi(s)$ and $\Phi(s)$ defined as

$$\varphi(s) = \sum_{n=1}^{\infty} \frac{\alpha_n}{\lambda_n^s}, \qquad \sigma > \sigma_\varphi, \tag{9.47}$$

$$\Phi(s) = \sum_{n=1}^{\infty} \frac{a_n}{\gamma_n^s}, \qquad \sigma > \sigma_\Phi, \tag{9.48}$$

where $\{\lambda_n\}$ and $\{\gamma_n\}$ are increasing sequences of real numbers and α_n and a_n are complex numbers. We assume that they are continued to meromorphic functions over the whole plane and that they satisfy the growth condition

$$\varphi(\sigma + it) << (|t| + 1)^{s_\varphi(\sigma)}, \quad \Phi(\sigma + it) << (|t| + 1)^{s_\Phi(\sigma)} \tag{9.49}$$

in the strip $-b < \sigma < c$. E.g. in the case of the Riemann zeta-function, $s_\zeta(-b) = \frac{1}{2} + b$. The latter works as the catalytic Dirichlet series as in [MR, §6.3.3]. Cf. the proof of Theorem 9.5, (ii).

The analytic continuation is most often supplied by the functional equation, cf. (7.8)

$$\varphi(s) = G(s)\psi(r - s), \tag{9.50}$$

which we assume in the last stage of analysis, where

$$G(s) = \frac{\Gamma(\{e_j + E_j(r - s)\}_{j=1}^N)}{\Gamma(\{f_j + F_j(r - s)\}_{j=1}^Q)} \frac{\Gamma(\{c_j + C_j s\}_{j=1}^P)}{\Gamma(\{d_j + D_j s\}_{j=1}^M)}. \tag{9.51}$$

In the beginning we consider φ and Φ just as Dirichlet series and form the integral ($c > 0$, $\varkappa \geq 0$)

$$\mathcal{F}_c(u, v) = \mathcal{F}_c^\varkappa(u, v; x) := \frac{1}{2\pi i} \int_{(c)} \frac{\Gamma(w)}{\Gamma(w + \varkappa + 1)} \varphi(u+w)\Phi(v-w)x^{w+\varkappa}\, dw, \tag{9.52}$$

and its counterpart $\mathcal{F}(v, u)$ under the condition

$$\operatorname{Re} u > \sigma_\varphi + c, \quad \operatorname{Re} v > \sigma_\Phi + c, \tag{9.53}$$

which assures the absolute convergence of the Dirichlet series.

Theorem 9.5. *(i) The Atkinson dissection is the special case of the Riesz sum $A^\varkappa(x)$ with $\varkappa = 0$ in the sense that*

$$\mathcal{F}(u,v)+\mathcal{F}(v,u)=\frac{1}{\Gamma(\varkappa+1)}\sum_{m=1}^{\infty}a_m\gamma_m^{-v-\varkappa}\sum_{\lambda_n\le\gamma_m x}{}' \alpha_n\lambda_n^{-u}(\gamma_m x-\lambda_n)^{\varkappa}$$
$$+\frac{1}{\Gamma(\varkappa+1)}\sum_{n=1}^{\infty}\alpha_n\lambda_n^{-u-\varkappa}\sum_{\gamma_m\le\lambda_n x}{}' a_m\gamma_m^{-v}(\lambda_n x-\gamma_m)^{\varkappa} \tag{9.54}$$

implies

$$\sum_{m,n=1}^{\infty}\alpha_m\lambda_m^{-u}a_n\gamma_n^{-v}=\sum_{m=1}^{\infty}\sum_{n<m}\alpha_m\lambda_m^{-u}a_n\gamma_n^{-v} \tag{9.55}$$
$$+\sum_{m=n}^{\infty}\alpha_n\lambda_n^{-u}a_n\gamma_n^{-v}+\sum_{n=1}^{\infty}\sum_{m<n}\alpha_m\lambda_m^{-u}a_n\gamma_n^{-v}.$$

(ii) Suppose for $b>0$

$$\operatorname{Re}u>\max\{\sigma_\varphi+c,s_\varphi(-b)-\varkappa\},\quad \operatorname{Re}v>\max\{\sigma_\Phi+c,s_\Phi(-b)-\varkappa\}. \tag{9.56}$$

Then

$$\mathcal{F}(u,v)+\mathcal{F}(v,u)=g(u,v)+g(v,u)+\mathrm{P}(x), \tag{9.57}$$

where

$$g(u,v):=\sum_{\ell_k=1}^{\infty}\frac{\tilde{\sigma}_{r-u-v}(\ell_k)}{\ell_k^{u-r}}\frac{1}{2\pi i}\int_{(-b)}\frac{\Gamma(w)}{\Gamma(w+\varkappa+1)}G(u+w)\left(\frac{1}{\ell_k x}\right)^{-w}\mathrm{d}w, \tag{9.58}$$

and where

$$\tilde{\sigma}_{r-u-v}(\ell_k)=\sum_{\lambda_n\gamma_m=\ell_k}a_m\alpha_n\gamma_m^{r-u-v} \tag{9.59}$$

and $\mathrm{P}(x)$ *indicates the sum of residues of the integrand in the strip* $-b<\sigma<c$.

(iii) We have the closed formula

$$g(u,v)=\mathcal{F}_{-b}(u,v)=\sum_{\ell_k=1}^{\infty}\frac{\tilde{\sigma}_{r-u-v}(\ell_k)}{\ell_k^{u-r}}$$
$$H_{M+1,Q}^{P+1,N}\left(\frac{1}{\ell_k x}\,\middle|\,\begin{matrix}\{(1-(e_j+E_j(r-u)),E_j)\}_{j=1}^N,(\varkappa+1,1),\{(d_j,D_j)\}_{j=1}^M\\(0,1),\{(c_j,C_j)\}_{j=1}^P,\{(1-(f_j+F_j(r-u)),F_j)\}_{j=1}^Q\end{matrix}\right). \tag{9.60}$$

Proof. (i) Substituting and changing the order of summation and integration, we obtain

$$\mathcal{F}(u,v) = \sum_{m=1}^{\infty} a_m \gamma_m^{-v-\varkappa} \sum_{n=1}^{\infty} \alpha_n \lambda_n^{-u} \frac{x^{\varkappa}}{2\pi i} \int_{(c)} \frac{\Gamma(w)}{\Gamma(w+\varkappa+1)} \left(\frac{\lambda_n}{\gamma_m x}\right)^{-w} dw, \tag{9.61}$$

whence

$$\mathcal{F}(u,v) = \frac{1}{\Gamma(\varkappa+1)} \sum_{m=1}^{\infty} a_m \gamma_m^{-u-\varkappa} \sideset{}{'}\sum_{\lambda_n \le \gamma_m x} \lambda_n^{-u} (\gamma_m x - \lambda_m)^{\varkappa} \alpha_n. \tag{9.62}$$

From (9.62) and its counterpart for $\mathcal{F}(v,u)$, we deduce (9.54).

The special case of (9.54) $\varkappa = 0, x = 1$ leads to (9.55) on account of the discontinuous integral.

(ii) To shift the line of integration to $\sigma = -b$, we must assure the (absolute) convergence of the resulting integrals by Condition (9.56). However, on the line $\sigma = -b$, the Dirichlet series still remains a Dirichlet series and is *inert in the order of magnitude.*

In order to apply the functional equation, we must have $r+b-\mathrm{Re}\, u > \sigma_\varphi$ which is one of the conditions in (9.56). But to make the range of consistent, we need to take $\varkappa > 0$. Under the condition (9.56), we may apply the functional equation (9.50) and we obtain (9.60).

This is a typical example of the $\varkappa$ times integration of the original sum and by executing $\varkappa$ times differentiation leads to the expression for the original sum. □

Example 9.3. In the case of the Riemann zeta-function, we want to have $\mathrm{Re}\, u < b$ as well as $\mathrm{Re}\, u > b + \frac{1}{2} - \varkappa$. Here is a conflict and we must have $\varkappa \ge 1$ rather than $\varkappa = 0$. (9.58) reads

$$g^{\varkappa}(u,v) = x^{\varkappa} \sum_{\ell=1}^{\infty} \frac{\sigma_{1-u-v}(\ell)}{\ell^{u-1}} H_{1,3}^{2,0}\left(\frac{1}{\ell x} \,\middle|\, \begin{matrix} \left(\frac{1+u}{2}, \frac{1}{2}\right), (\varkappa+1, 1), \left(\frac{u}{2}, \frac{1}{2}\right) \\ (0,1), \end{matrix}\right). \tag{9.63}$$

By the standard technique of augmentation, duplication and reciprocation, we may transform the H-function in (9.63) into

$$H_{1,3}^{2,0} = \frac{2^{u-1}}{\sqrt{\pi i}} \left(e^{\frac{\pi i}{2} u} G_{2,1}^{1,1}\left(-\frac{i}{2\ell x} \,\middle|\, \begin{matrix} u, \varkappa+1 \\ 0 \end{matrix}\right) + e^{-\frac{\pi i}{2} u} G_{2,1}^{1,1}\left(\frac{i}{2\ell x} \,\middle|\, \begin{matrix} u, \varkappa+1 \\ 0 \end{matrix}\right)\right), \tag{9.64}$$

whence by inversion

$$H_{1,3}^{2,0} = \frac{2^{u-1}}{\sqrt{\pi}i}\left(e^{\frac{\pi i}{2}u}G_{1,2}^{1,1}\left(2i\ell x\,\middle|\,\begin{matrix}1\\1-u,-\varkappa\end{matrix}\right)+e^{-\frac{\pi i}{2}u}G_{1,2}^{1,1}\left(-2i\ell x\,\middle|\,\begin{matrix}1\\1-u,-\varkappa\end{matrix}\right)\right). \tag{9.65}$$

On the other hand, the residual function is apparently

$$\mathrm{P}_1(x) = 2\zeta(u)\zeta(v)x - \zeta(u+v-1)\left(\frac{1}{u(u-1)}x^u + \frac{1}{v(v-1)}x^v\right). \tag{9.66}$$

Hence differentiation leads to

$$\mathrm{P}(x) = 2\zeta(u)\zeta(v) - \zeta(u+v-1)\left(\frac{1}{u-1}x^{u-1} + \frac{1}{v-1}x^{v-1}\right). \tag{9.67}$$

Theorem 9.6. (Generalization of Wilton's approximate formula)

$$\mathcal{F}_c(u,v) + \mathcal{F}_c(v,u) = g(u,v) + g(v,u) + \mathrm{P}_1(x), \tag{9.68}$$

where $\mathrm{P}_1(x)$ *is given by* (9.66) *and*

$$\begin{aligned} g^{\varkappa}(u,v) &= \frac{i}{\sqrt{\pi}}x^{\varkappa+1-u}\frac{\Gamma(1-u)}{\Gamma(2+\varkappa-u)}\sum_{\ell=1}^{\infty}\sigma_{1-u-v}(\ell) \\ &\times\left({}_1F_1\left(\begin{matrix}1-u\\2+\varkappa-u\end{matrix};-2i\ell x\right) - {}_1F_1\left(\begin{matrix}1-u\\2+\varkappa-u\end{matrix};2i\ell x\right)\right). \end{aligned} \tag{9.69}$$

To deduce Wilton's formula, we may simply take $\varkappa = 1$ and differentiate both sides of (9.57) with respect to x.

Since $g(u,v)$ is the sum of two terms of the form $zG_{1,2}^{1,1}\left(z\,\middle|\,\begin{matrix}1\\1-u,-1\end{matrix}\right)$ and

$$\frac{\mathrm{d}}{\mathrm{d}z}zG_{1,2}^{1,1}\left(z\,\middle|\,\begin{matrix}1\\1-u,-1\end{matrix}\right) = -G_{1,1}^{1,2}\left(z\,\middle|\,\begin{matrix}1\\1-u,0\end{matrix}\right) = \Gamma(1-u,z) - \Gamma(1-u), \tag{9.70}$$

we see that we have the incomplete gamma functions on the right-hand side.

On other hand, differentiation of $\mathcal{F}_c^1(u,v)$ by (9.1) amounts to $\zeta(u)\zeta(v)$, which cancels one term in (9.67) and we obtain *Wilton's approximate functional equation*

Corollary 9.3. (Wilton) *For* $\mathrm{Re}\,u > -1$, $\mathrm{Re}\,v > -1$, $\mathrm{Re}(u+v) > -1$, $u+v \neq 2$

$$\begin{aligned}\zeta(u)\zeta(v) &= \zeta(u+v-1)\left(\frac{1}{u-1}+\frac{1}{v-1}\right)\\ &+ 2(2\pi)^{u-1}\sum_{\ell=1}^{\infty}\frac{\sigma_{1-u-v}(\ell)}{\ell^{u-1}}u\int_{2\pi\ell}^{\infty}x^{-u-1}\sin x\,\mathrm{d}x \\ &+ 2(2\pi)^{v-1}\sum_{\ell=1}^{\infty}\frac{\sigma_{1-u-v}(\ell)}{\ell^{v-1}}v\int_{2\pi\ell}^{\infty}x^{-v-1}\sin x\,\mathrm{d}x.\end{aligned} \tag{9.71}$$

9.5. *The product of Hurwitz zeta-functions*

This subsection partly depends on [WL], which elucidates and generalizes the result of Nakajima [Nak1] on the product of two Hurwitz zeta-functions. Since the Hurwitz zeta-function satisfies a ramified but of Hecke type functional equation, the analysis is much simpler and the main ingredient is the K-Bessel function. With this the general case can be worked with.

We define for $\boldsymbol{\alpha} = (\alpha_1, \dots, \alpha_\varkappa) \in \mathbb{R}^\varkappa$ the summatory function of the $\varkappa$-dimensional shifted divisor function

$$D(x;\boldsymbol{\alpha}) = \sideset{}{'}\sum_{\substack{(m_1+\alpha_1)\cdots(m_\varkappa+\alpha_\varkappa)\le x\\ m_j\in\mathbb{N}\cup\{0\}}} 1 \tag{9.72}$$

and the $\varkappa$-dimensional divisor sum

$$d_{\boldsymbol{\alpha}}(n) = d_{\alpha_1,\cdots\alpha_\varkappa}(n) = \sum_{\substack{m_j\in\mathbb{N}\\ m_1\cdots m_\varkappa=n}} e^{2\pi i(\alpha_1 m_1+\cdots+\alpha_\varkappa m_\varkappa)}, \tag{9.73}$$

and we use a short-hand $d_{+(\varkappa-r),-r}(n)$ to indicate $d_{\boldsymbol{\alpha}}(n)$ with $(\varkappa-r)$ α_i's are positive and (r) α_i's are negative, e.g.

$$d_{+2,-1} = d_{+2,-1}(n) = \sum_{\substack{m_j\in\mathbb{N}\\ m_1m_2m_3=n}} e^{2\pi i(\alpha_1m_1+\alpha_2m_2-\alpha_3m_3)}.$$

Hence the result is the sum of the following.

$$\begin{aligned} d_{+\varkappa,-0}(n)\Big((2\pi)^\varkappa\left(e^{-\frac{\pi i}{2}}\right)^\varkappa x\Big)^s + d_{+(\varkappa-1),-1}(n)\bigg((2\pi)^\varkappa\left(e^{-\frac{\pi i}{2}}\right)^{\varkappa-2}x\bigg)^s + \cdots \\ + d_{+0,-\varkappa,-0}(n)\Big((2\pi)^\varkappa\left(e^{\frac{\pi i}{2}}\right)^\varkappa x\Big)^s + d_{+1,-(\varkappa-1)}(n)\bigg((2\pi)^\varkappa\left(e^{\frac{\pi i}{2}}\right)^{\varkappa-2}x\bigg)^s. \end{aligned} \tag{9.74}$$

Table 3. Summands

$\varkappa$	$+\varkappa,-0$	$+(\varkappa-1),-1$	$\cdot$	$\pm\frac{\varkappa}{2}$
$\varkappa$	$d_{+\varkappa,-0}(n)\left(e^{-\frac{\pi i}{2}s}\right)^{\varkappa}$	$d_{+(\varkappa-1),-1}(n)\left(e^{-\frac{\pi i}{2}s}\right)^{\varkappa-2}$	$\cdot$	$d_{\pm\frac{\varkappa}{2}}$
$\varkappa$	$d_{+0,-\varkappa}(n)\left(e^{\frac{\pi i}{2}s}\right)^{\varkappa}$	$d_{+1,-(\varkappa-1)}(n)\left(e^{-\frac{\pi i}{2}s}\right)^{\varkappa-2}$	$\cdot$	$-$
$4k$	$d_{+4k,-0}(n)$	$d_{+(4k-1),-1}(n)e^{\pi is}$	$\cdot$	$d_{\pm(2k)}$
$4k$	$d_{+0,-4k}(n)$	$d_{+1,-(4k-1)}(n)e^{-\pi is}$	$\cdot$	$-$
$4k+1$	$d_{+(4k+1),-0}(n)e^{-\frac{\pi i}{2}s}$	$d_{+(4k),-1}(n)e^{\frac{\pi i}{2}s}$	$\cdot$	$-$
$4k+1$	$d_{+0,-(4k+1)}(n)e^{\frac{\pi i}{2}s}$	$d_{+1,-(4k)}(n)e^{-\frac{\pi i}{2}s}$	$\cdot$	$-$
$4k+2$	$d_{+(4k+2),-0}(n)e^{-\pi is}$	$d_{+(4k+1),-1}(n)$	$\cdot$	$d_{\pm(2k+1)}$
$4k+2$	$d_{+0,-(4k+2)}(n)e^{\pi is}$	$d_{+1,-(4k+1)}(n)$	$\cdot$	$-$
$4k+3$	$d_{+(4k+3),-0}(n)e^{-\frac{3\pi i}{2}s}$	$d_{+(4k+2),-1}(n)e^{-\frac{\pi i}{2}s}$	$\cdot$	$-$
$4k+3$	$d_{+0,-(4k+3)}(n)e^{\frac{3\pi i}{2}s}$	$d_{+1,-(4k+2)}(n)e^{\frac{\pi i}{2}s}$	$\cdot$	$-$
$\varkappa=3$	$d_{+3,-0}(n)e^{-\frac{3\pi i}{2}s}$	$d_{+2,-1}(n)e^{-\frac{\pi i}{2}s}$	$\cdot$	$-$
$\varkappa=3$	$d_{+0,-3}(n)e^{\frac{3\pi i}{2}s}$	$d_{+1,-2}(n)e^{\frac{\pi i}{2}s}$	$\cdot$	$-$

Theorem 9.7. *For* $\boldsymbol{\alpha}=(\alpha_1,\cdots\alpha_\varkappa)\in(0,1)^\varkappa$, *we have*

$$\begin{aligned} D(x;\boldsymbol{\alpha}) &= \mathrm{P}(x) \\ &+\sum_{n=1}^{\infty}\left(d_{+\varkappa,-0}(n)V((2\pi)^\varkappa\left(e^{-\frac{\pi i}{2}}\right)^\varkappa x\middle|1,\cdots,1,0)\right. \\ &\left.+\,d_{+(\varkappa-1),-1}(n)V((2\pi)^\varkappa\left(e^{-\frac{\pi i}{2}}\right)^{\varkappa-2}x\middle|1,\cdots,1,0\right). \end{aligned} \tag{9.75}$$

Corollary 9.4. *(i) For* $\boldsymbol{\alpha}=(\alpha_1,\alpha_2)\in(0,1)^2$, *we have*

$$\begin{aligned} &D(x;\boldsymbol{\alpha})=\mathrm{P}(x) \\ &+\sum_{n=1}^{\infty}\left(d_{\alpha_1,\alpha_2}(n)V\left((2\pi)^2e^{-\pi i}x\middle|1,0\right)+\left(d_{\alpha_1,-\alpha_2}(n)+d_{-\alpha_1,\alpha_2}(n)\right)\right. \\ &\cdot V\left((2\pi)^2x\middle|1,0\right) \\ &\left.+\,d_{-\alpha_1,-\alpha_2}(n)V\left((2\pi)^2e^{\pi i}x\middle|1,0\right)\right), \end{aligned} \tag{9.76}$$

where

$$\mathrm{P}(x) = \mathrm{P}_{\alpha_1,\alpha_2}(x) = x\log x + \left\{\left(-\frac{\Gamma'}{\Gamma}(\alpha_1)\right) + \left(-\frac{\Gamma'}{\Gamma}(\alpha_2)\right) - 1\right\}x + \left(\frac{1}{2}-\alpha_1\right)\left(\frac{1}{2}-\alpha_2\right),$$

(ii) For $\boldsymbol{\alpha} = (\alpha_1, \alpha_2, \alpha_3) \in (0,1)^3$, *we have*

$$\begin{aligned} D(x;\boldsymbol{\alpha}) &= \mathrm{P}(x) \\ &+ \sum_{n=1}^{\infty}\Bigg(d_{+3,-0}(n)V\left((2\pi)^3\left(e^{-\frac{\pi i}{2}}\right)^3 x\,\middle|\,1,1,0\right) \\ &+ d_{+2,-1}(n)V\left((2\pi)^3 e^{-\frac{\pi i}{2}}x\,\middle|\,1,1,0\right) + d_{+1,-2}(n)V\left((2\pi)^3 e^{-\frac{\pi i}{2}}x\,\middle|\,1,1,0\right) \\ &+ d_{+0,-3}(n)V\left((2\pi)^3\left(e^{\frac{\pi i}{2}}\right)^3 x\,\middle|\,1,1,0\right)\Bigg), \end{aligned} \tag{9.77}$$

where there are three of $d_{+2,-1}(n)$ *and* $d_{+1,-2}(n)$, *respectively and where* $\mathrm{P}(x)$ *is the sum of the residues of the integrand* $\zeta(s,\alpha_1)\zeta(s,\alpha_2)\zeta(s,\alpha_3)\frac{x^s}{s}$ *at* $s = 0, 1$.

References

ApoMA. T. M. Apostol, *Mathematical analysis*, Addison-Wesley, Reading, 1957.

Atk2. F.V. Atkinson, The mean value of the Riemann zeta-function, *Acta Math.* **81** (1949), 353–376.

Bel1. R. Bellman, An analog of an identity due to Wilton, *Duke Math. J.* **16** (1949), 539–545.

Bel2. R. Bellman, Wigert's approximate functional equation and the Riemann zeta-function, *Duke Math. J.* **16** (1949), 547–552.

Bel3. R. Bellman, Generalized Eisenstein series and non-analytic automorphic functions, *Proc. Nat. Acad. Sci., USA* **36** (1950), 356–359.

Bel4. R. Bellman, On the functional equations of the Dirichlet series derived from Siegel modular forms, *Proc. Nat. Acad. Sci., USA* **37** (1951), 84–87.

Bel5. [Bel5] R. Bellman, On a class of functional equations of modular type, *Proc. Nat. Acad. Sci., USA* **42** (1956), 84–87.

BeI. B. C. Berndt, Identities involving the coefficients of a class of Dirichlet series I, *Trans. Amer. Math. Soc.* **137** (1969), 345–359.

BeII. B. C. Berndt, Identities involving the coefficients of a class of Dirichlet series II, *Trans. Amer. Math. Soc.* **137** (1969), 361–374.

Vista II. K. Chakraborty, S. Kanemitsu and H. Tsukada, Vistas of special functions II, World Sci.. New Jersey-London-Singapore etc. 2009.

FB. FB K. Chakraborty, S. Kanemitsu and H. Tsukada, Applications of the beta transform, to appear.

Ch. K. Chandrasekharan, *Arithmetical Functions*, Springer Verl., Berlin-New York etc. 1969.

CJ. K. Chandrasekharan and H. Joris, Dirichlet series with functional equations and related arithmetical functions, *Acta Arith.* **24** (1973), 165–191.

ChTyp. [ChTyp] K. Chandrasekharan and S. Minakshisundaram, *Typical means*, Oxford UP, Oxford 1952.

ChN3. K. Chandrasekharan and Raghavan Narasimhan, Functional equations with multiple gamma factors and the average order of arithmetical functions. *Ann. of Math.* (2) **76** (1962), 93–136.

CS. S. Chowla and A. Selberg, On Epstein's zeta-function (I), *Proc. Nat. Acad. Sci. USA* **35** (1949), 371–374; *Collected Papers of Atle Selberg I*, Springer Verlag, 1989, 367–370. *The Collected Papers of Sarvadaman Chowla II*, CRM, 1999, 719–722.

Coh. H. Cohen, q-identities for Maass waveforms, *Invent. Math.* **91**(1998), 409-422.

Dav. H. Davenport, Multiplicative number theory, 1st ed. Markham, Chicago 1967, 2nd ed. Springer, New York etc. 1980.

DeM. Debika Banerjee and Jay Mehta, Linearized product of zeta-functions, Proc. Japan Acad. Ser. A Math., to appear.

ErdH. A. Erdélyi, W. Magnus, F. Oberhettinger and F. G. Tricomi, *Higher transcendental functions*, Vols I-III, Based, in part, on notes left by Harry Bateman, McGraw-Hill, New York 1953.

ErdT. A. Erdélyi, W. Magnus, F. Oberhettinger and F. G. Tricomi, *Tables of Integral Transforms. Vols. I-II* Based, in part, on notes left by Harry Bateman, McGraw-Hill Book Company, Inc., New York-Toronto-London, 1954.

Ham2. H. Hamburger, Über einige Beziehungen, die mit der Funktionalgleichung der Riemannschen ζ-Funktion äquivalent sind, *Math. Ann.* **85** (1922), 129-140.

HarMB. G. H. Hardy, Some multiple integrals, *Quart. J. Math. (Oxford)* (2) **5** (1908), 357–375; *Collected Papers. Vol. V* (1972), 434–452, Comments 453.

HarRie. G. H. Hardy and M. Riesz, *The general theory of Dirichlet's series*, CUP. Cambridge 1915; reprint, Hafner, New York 1972.

KaR. [KaR] S. Kanemitsu, On the Riesz sums of some arithmetical functions, *in p-adic L-functions and algebraic number thery Surikaiseki Kenkyusho Kokyuroku* **411** (1981), 109-120.

CZS3. S. Kanemitsu, Y. Tanigawa, H. Tsukada and M. Yoshimoto, Crystal symmetry viewed as zeta symmetry, *Proceedings of Kinki University Symposium "Zeta Functions, Topology and Quantum Physics 2003"* (Developments in Mathematics, Vol.14) 91–129, Springer Verl., Berlin etc. 2005.

KTY7. S. Kanemitsu, Y. Tanigawa and M. Yoshimoto, Ramanujan's formula and modular forms, *Number-theoretic Methods – Future Trends, Proceedings of a conference held in Iizuka* (S. Kanemitsu and C. Jia, eds.), Kluwer, Dordrecht, 2002, pp. 159–212.

Vista I. S. Kanemitsu and H. Tsukada, Vistas of special functions, World Sci.. New Jersey-London-Singapore etc. 2007.

MR. S. Kanemitsu and H. Tsukada, *Contributions to the theory of zeta-functions: the modular relation supremacy*, World Scientific, Singapore-London-New York. 2014.

Kat1. M. Katsurada, On Mellin-Barnes type of integrals and sums associated with the Riemann zeta-function, *Publ. Inst. Math.* **62** (1997), 13–25.

Kat2. M. Katsurada, Power series and asymptotic series associated with the Lerch zeta-function, *Proc. Japan Acad. Ser. A* **74** (1998), 167-170.

Kob2. H. Kober, Eine der Riemannschen verwandte Funktionalgleichung, *Math. Z.* **39** (1935), 630-633.

Kosum. N. S. Koshlyakov, Application of the theory of sum-formulae to the investigation of a class of one-valued analytical functions in the theory of numbers, *Mess. Math.* **58** (1928/29), 1-23.

Kosvor. N. S. Koshlyakov, On Voronoǐ's sum formula, *Mess. Math.* **58** (1928/29), 30-32.

KoshI. N. S. Koshlyakov, Investigation of some questions of analytic theory of the rational and quadratic fields, I (Russian), *Izv. Akad. Nauk SSSR, Ser. Mat.* **18** (1954), 113–144.

KoshII. N. S. Koshlyakov, Investigation of some questions of analytic theory of the rational and quadratic fields, II (Russian), *Izv. Akad. Nauk SSSR, Ser. Mat.* **18** (1954), 213–260, Errata: ibid. **19** (1955), 271 (in Russian).

KoshIII. N. S. Koshlyakov, Investigation of some questions of analytic theory of the rational and quadratic fields, III (Russian), *Izv. Akad. Nauk SSSR, Ser. Mat.* **18** (1954), 307–326, Errata: ibid. **19** (1955), 271 (in Russian).

KoshE. N. S. Koshlyakov, Letter to the editor, *Izv. Akad. Nauk SSSR, Ser. Mat.* **19** (1955), 271 (in Russian).

LanII. E. Landau, Über die Anzahl der Gitterpunkte in gewissen Bereichen (Zweite Mit.), *Nachr. Ges. Wiss. Göttingen, Math.-Phys. Kl.* (1915), 209-243=Collected Works Vol. 6, Thales Verl., Essen 1985, 308-342.

LanZT. E. Landau, Vorlesungen über Zahlentheorie, Teubner, Leipzig 1921 and reprint Chelsea, New York 1967.

LKT. H.-L.Li, S. Kanemitsu and H. Tsukada, Modular relation interpretation of the series involving the Riemann zeta values, *Proc. Japan Acad. Ser. A*, **84** (2008), 154-158.

LWKman. F.-H.Li, N.-L. Wang and S. Kanemitsu, Manifestations of the general modular relation, *Šiaulai Math. Sem.* **7** (2012), 59-77.

LWK. F.-H.Li, N.-L. Wang and S. Kanemitsu, Number theory and its applications, World Scientific, Singapore etc. 2013.

MP. F. Mainardi and G. Pagnini, Salvatore Pincherle: the pioneer of the Mellin-Barnes integrals, J. Comp. Appl. Math. (2003), 331-342.

Nak1. M. Nakajima, Shifted divisor problem and random divisor problem. *Proc. Japan, Acad. Ser. A Math..* **69** (1993), 49-52.

Nak2. M. Nakajima, A new expression for the product of two Dirichlet series, *Proc. Japan Acad. Ser. A Math.* **79** (2003), 49-52.

PBM. A. P. Prudnikov, Yu. A. Bychkov and O. I. Marichev, *Integrals and Series, Supplementary Chapters*, Izd. Nauka, Moscow 1986.

Sa. F. Sato, Searching for the origin of prehomogeneous vector spaces, at annual meeting of the Math. Soc. Japan 1992 (in Japanese).

Seg. S. L. Segal, On $\sum 1/n \sin(x/n)$ and $\sum 1/n \cos(x/n)$, J. London Math. Soc. (2) **4** (1972), 385-393.

SC. A. Selberg and S.Chowla, On Epstein's zeta-function, *J. Reine Angew, Math.* **227** (1967), 86–110; *Collected Papers of Atle Selberg I*, Springer Verlag, 1989, 521–545; *The Collected Papers of Sarvadaman Chowla II*, CRM, 1999, 1101–1125.

Siegl. C. L. Siegel, Bemerkungen zu einem Satz von Hamburger über die Funktionalgleichung der Riemannschen Zetafunktion, Math. Ann. **85** (1922), 276-279=Ges. Abh., Bd. I, Springer Verl. Berlin-Heidelberg, 1966, 154-156.

Sie3. C. L. Siegel, Über die Zetafunktionen indefiniter quadratische Formen, I, II *Math. Z.* **43** (1938), 682–708; **44** (1939), 398–426 =*Ges. Abh.*, Bd. II, Springer Verl. Berlin-Heidelberg, 1966, 41-67; 68-96.

Sie4. C. L. Siegel, Indefinite quadratische Formen und Funktionentheorie, I, *Math. Ann.* **124** (1951), 17–54=*Ges. Abh.*, Bd. III, Springer Verl. Berlin-Heidelberg, 1966, 105-142.

Spl. W. Splettstösser, Some aspects of the reconstruction of sampled signal functions, 126-142.

St. S. W. P. Steen, Divisor functions: their differential equations and recurrence formulas, *Proc. London Math. Soc.* (2) **31** (1930), 47–80.

Te5. A. Terras, *Harmonic analysis on symmetric spaces and applications I, II*, Springer Verl., New York etc. 1985.

Ts. H. Tsukada, A general modular relation in analytic number theory. *Number Theory: Sailing on the sea of number theory, Proc. 4th China-Japan Seminar on number theory 2006* 214–236, World Sci., Singapore etc. 2007.

Vor1. G. F. Voronoï, Sur une fonction transcendente et ses applications à la sommation de qulques séries, Ann. École Norm. Sup. (3) **21** (1904), 459-533=Sob. Soč. II, Izd. Akad. Nauk Ukr. SSR, Kiev 1952, 51-165.

AAW1. A. A. Walfisz, On sums of coefficients of some Dirichlet series, *Soobšč. Akad. Nauk Grundz. SSR* **26** (1961), 9–16.

AAW2. A. A. Walfisz, On the theory of a class of Dirichlet series, *ibid.* **27** (1961), 9–16.

AAW4. A. A. Walfisz, The Fourier-Poisson formula a class of Dirichlet series, *Trudy Tbilis. Mat. Inst. im. A. M. Razmadze* **29** (1963), 1–13.

WL. N.-L. Wang and H.-L. Li, On the product of Hurwitz zeta-functions, to appear.

WW. X.-H. Wang and N.-L. Wang, Modular-relation-theoretic interpretation of M. Katsurada's results, to appear.

Wea. H. J. Weaver, Applications of discrete and continuous Fourier transforms, Wiley, New York etc. 1983.

Wil4. J. R. Wilton, An extended form of Dirichlet's divisor problem, *Proc. London Math. Soc.* (2) **36** (1933), 391–426.

Wil2. J. R. Wilton, An approximate functional equation for the product of two ζ-functions, *Proc. London Math. Soc.* (2) **31** (1930), 11–17.

Wis. J. Wishart, The generalized product moment distributions in samples from a normal multiplicative population, *Biometrika* **20** (1928), 32-43.

FIGURATE PRIMES AND HILBERT'S 8TH PROBLEM

TIANXIN CAI
Department of Mathematics, Zhejiang University,
Hangzhou, 310027, People's Republic of China
E-mail: txcai@zju.edu.cn

YONG ZHANG
Department of Mathematics, Zhejiang University,
Hangzhou, 310027, People's Republic of China
E-mail: zhangyongzju@163.com

ZHONGYAN SHEN
Department of Mathematics, Zhejiang International Study University,
Hangzhou, 310012, People's Republic of China
E-mail: huanchenszyan@yahoo.com.cn

1. Figurate primes

In a letter to the editor of Crelle's Journal in 1844, Catalan stated that 8 and 9 are the only consecutive perfect powers, i.e., the Diophantine equation

$$p^a - q^b = 1$$

has unique positive integral solution $(p, q, a, b) = (2, 3, 3, 2)$. This is later known as Catalan's conjecture.

In 2004, P. Mihăilescu [4] proved this conjecture by making extensive use of the theory of cyclotomic fields and Galois modules.

More generally, we have the Diophantine equation

$$p^a - q^b = k, \tag{1.1}$$

where p, q are primes, $a, b, k \in \mathbb{N}$.

When $a = 1, q = 2, k = 1$, the solutions of (1.1) are exactly Fermat primes.

When $p = 2, b = 1, k = 1$, the solutions of (1.1) are exactly Mesernne primes.

When $a, b \geq 2, k = 1$, (1.1) is the Diophantine equation for Catalan's conjecture.

There are many authors who have investigated this problem for $k > 1$ and more information can be found in [3]: D9 Catalan conjecture & Difference of two powers and D10 Exponential diophantine equations.

When $a = b = 1, k = 2$, (1.1) is the Diophantine equation for twin primes conjecture.

In spring 2013, the first author [2] defined figurate primes as the positive binomial coefficients

$$\binom{p^a}{i}, \ a \geq 1, \ i \geq 1,$$

where p is a prime. The set of figurate primes includes all the primes, but its density among positive integers is the same as the set of primes. We study the following Diophantine equation

$$\binom{p^a}{i} - \binom{q^b}{j} = k, \tag{1.2}$$

where p, q are primes, $a, b, i, j, k \in \mathbb{N}$.

When $k = 1$, for $j = 1, i \geq 2$, we use an elementary method to prove

Theorem 1.1. *For $(i, j) = (2, 1)$, (1.2) has exactly four solutions $(p, q, a, b) = (2, 5, 2, 1), (3, 2, 1, 1), (2, 3, 3, 3), (5, 3, 1, 2)$. For $(i, j) = (3, 1)$, (1.2) has exactly three solutions $(p, q, a, b) = (2, 3, 2, 1), (3, 83, 2, 1), (5, 3, 1, 2)$. For $(i, j) = (4, 1)$, (1.2) has exactly two solutions $(p, q, a, b) = (5, 2, 1, 2), (3, 5, 2, 3)$.*

For $i = b = 1, j = 2$. If a is even, it is easy to see that (1.2) has unique solution $(p, q) = (2, 3)$; if $a = 1$, it seems likely that (1.2) has infinitely many solutions, i.e., there are infinitely many pairs of primes (p, q) satisfying

$$p - 1 = \binom{q}{2}.$$

However, it is even a harder problem than that of prime representations by binary forms. The least 10 examples are $(p, q) = (2, 2)$, $(11, 5)$, $(79, 13)$, $(137, 17)$, $(821, 41)$, $(1831, 61)$, $(3917, 89)$, $(4657, 97)$, $(5051, 101)$, $(6329, 113)$; if $a > 1$ is odd, we guess that (1.2) has no solutions. It is true for $a = 3$ by an easy calculation.

Similarly, as in the proof of Theorem 1.1, we may find that all the solutions of the Diophantine equation

$$p^a - 1 = \binom{q^b}{3}$$

are $(p, q, a, b) = (5, 2, 1, 2), (2, 3, 1, 1), (11, 5, 1, 1)$.

For $i = j \geq 2$, it is easy to verify that (1.2) has no solution. By using the theory of elliptic curves and **Magma**, we have

Theorem 1.2. *For $(i, j) = (2, 3)$, (1.2) has unique solution $(p, q, a, b) = (3, 7, 2, 1)$; and for $(i, j) = (3, 2)$, (1.2) has exactly two solutions $(p, q, a, b) = (2, 3, 2, 1), (3, 7, 2, 1)$. For $(i, j) = (2, 4)$, (1.2) has exactly two solutions $(p, q, a, b) = (2, 5, 2, 1), (3, 7, 2, 1)$; and for $(i, j) = (4, 2)$, (1.2) has no solutions.*

When $k = 2$, we have similarly

Theorem 1.3. *For $(i, j) = (2, 3)$, (1.2) has exactly two solutions $(p, q, a, b) = (2, 2, 2, 2), (3, 3, 1, 1)$; and for $(i, j) = (3, 2)$, (1.2) has no solutions. For $(i, j) = (2, 4)$, (1.2) has unique solutions $(p, q, a, b) = (3, 2, 1, 2)$; and for $(i, j) = (4, 2)$, (1.2) has unique solution $(p, q, a, b) = (5, 3, 1, 1)$.*

2. Hilbert's 8th problem

Among the 23 problems that David Hilbert raised at the International Congress of Mathematicians in Paris in 1900, the 8th one might be the most profound and difficult, as it includes the Riemann Hypothesis, Goldbach's conjecture and twin primes conjecture.

Goldbach's conjecture (1742) is one of the most important unsolved problems in number theory:

Every even integer greater than 2 can be expressed as the sum of two primes, i.e.,

$$n = p + q, \ n \geq 4,$$

where n is even and p, q are primes. And every odd integer greater than 5 can be expressed as the sum of three primes, i.e.,

$$n = p + q + r, \ n \geq 7,$$

where n is odd and p, q, r are primes. The first half for even numbers is still an open problem.

However, by fundamental theorem of arithmetic, each positive integer can be constructed as the product of primes, prime numbers being the basic building blocks of any positive integer in multiplication. On the other hand, primes do not seem to play a key role in addition. Besides, it is not a perfect result that each even integer is the sum of two primes while each odd integer is the sum of three primes, under the validity of Goldbach's conjecture.

What we have to point out is that, among the composites in figurate primes the amount of even integers is as many as that of odd integers, they are more than powers of 2, cf. [3]: A19 Values of n making $n - 2^k$ prime & Odd numbers not of the form $\pm p^a \pm 2^b$. By calculations with computer, we check that every positive integer $1 < n \leq 10^7$ can be expressed as the sum of two figurate primes, i.e., the Diophantine equation

$$n = \binom{p^a}{i} + \binom{q^b}{j} \tag{2.1}$$

always has solutions with primes p, q and $a, b, i, j \in \mathbb{Z}^+$. With this support, we raise the following

Conjecture 2.1. *Every positive integer $n > 1$ can be expressed as the sum of two figurate primes.*

Conjecture 2.2. (*stronger than twin primes conjecture*) *There are infinitely many pairs of consecutive figurate primes.*

Moreover, if we read Hilbert's speech carefully, in the statement of the 8th problem, he mentioned

> *"After an exhaustive discussion of Riemann's prime number formula, perhaps we may sometime be in a position to attempt the rigorous solution of Goldbach's problem, viz., whether every integer is expressible as the sum of two positive prime numbers; and further to attack the well-known question, whether there are an infinite number of pairs of prime numbers with the difference 2, or even the more general problem, whether the linear diophantine equation*
>
> $$ax + by + c = 0$$
>
> (*with given integral coefficients each prime to the others*) *is always solvable in prime numbers x and y."*

With the idea of figurate primes and by numerical calculations, we have

Conjecture 2.3. *For any positive integers a and b, $(a, b) = 1$, when $n \geq (a-1)(b-1)$, there always exists a prime pair (x, y), such that*

$$ax + by = n.$$

Meanwhile, if we call a positive integer n a proper figurate prime if n is a figurate prime but not a prime. Then we even have a stronger

Conjecture 2.4. *Every positive integer $n > 5$ can be expressed as the sum of a prime and a proper figurate prime.*

3. Proof of the theorems

Proof of Theorem 1.1. When $k = 1$, for $(i, j) = (2, 1)$, (1.2) is equal to

$$(p^a + 1)(p^a - 2) = 2q^b.$$

Clearly, $d = (p^a + 1, p^a - 2) = 1$ or 3.

If $d = 1, p = 2$, we have

$$2^a - 2 = 2, 2^a + 1 = q^b$$

or

$$2^a - 2 = 2q^b, 2^a + 1 = 1,$$

it is easy to see that $(p, q, a, b) = (2, 5, 2, 1)$.

If $d = 1, p > 2$, we have

$$p^a - 2 = q^b, p^a + 1 = 2$$

or

$$p^a - 2 = 1, p^a + 1 = 2q^b,$$

it is easy to see that $(p, q, a, b) = (3, 2, 1, 1)$.

If $d = 3$, then $q = 3$. For $p = 2$, we have

$$2^a - 2 = 6, 2^a + 1 = 3^{b-1}$$

or

$$2^a - 2 = 2 \cdot 3^{b-1}, 2^a + 1 = 3,$$

it is easy to see that $(p, q, a, b) = (2, 3, 3, 3)$.

For $p > 2$, we have

$$p^a - 2 = 3, p^a + 1 = 2 \cdot 3^{b-1}$$

or

$$p^a - 2 = 3^{b-1}, p^a + 1 = 6,$$

it is easy to see that $(p, q, a, b) = (5, 3, 1, 2)$.

We can treat the cases $(i,j) = (3,1)$ or $(4,1)$ by a similar method. □

In the following proofs we use the notation $(X, \pm Y) = (10, -11; 189)$ to mean $(X, \pm Y) = (10, 189), (-11, 189)$, i.e. after the semi-colon, the y-coordinate follows.

Proof of Theorem 1.2. When $k = 1$, for convenience, put $p^a = y, q^b = x$ for $(i,j) = (2,3)$ and $p^a = x, q^b = y$ for $(i,j) = (3,2)$ in (1.2), respectively. Let

$$x = \frac{X + 12}{12}, \ y = \frac{Y + 36}{72},$$

the converse transformation is

$$X = 12x - 12, \ Y = 36(2y - 1).$$

Then, we have

$$Y^2 = X^3 - 144X + 11664, \ Y^2 = X^3 - 144X - 3024,$$

respectively.

Using **Magma**, we get all the integral points on the above two elliptic curves. For $Y^2 = X^3 - 144X + 11664$, the point $(X,Y) = (72, 612)$ leads to the unique solution $(p,q,a,b) = (3,7,2,1)$ of (1.2). For $Y^2 = X^3 - 144X - 3024$, the points $(X,Y) = (36,180), (84,756)$ lead to the two solutions $(p,q,a,b) = (2,3,2,1), (3,7,2,1)$ of (1.2).

Let $p^a = y, q^b = x$ for $(i,j) = (2,4)$ in (1.2), let

$$x = \frac{X + 3}{6}, \ y = Y + 2,$$

we have

$$Y^2 = 3X^4 + 6X^3 - 3X^2 - 6X + 81.$$

Using **Magma**, we get all the integral points on this elliptic curve, which are

$$(X, \pm Y) = (10, -11; 189), (-1, -2, 1, 0; 9), (-4, 3; 21), (-6, 5; 51),$$
$$(-92, 91; 14499).$$

We find that $(X,Y) = (3,21), (5,51)$ lead to the two solutions $(p,q,a,b) = (2,5,2,1), (3,7,2,1)$ of (1.2).

Let $p^a = x, q^b = y$ for $(i,j) = (4,2)$ in (1.2) and let

$$x = \frac{X + 3}{6}, \ y = Y + 2.$$

Then we have

$$Y^2 = 3X^4 + 6X^3 - 3X^2 - 6X - 63.$$

Using **Magma**, we get all the integral points on this elliptic curve, which are

$$(X, \pm Y) = (-3, 2; 3),$$

it is easy to see that there are no solution for $(i, j) = (4, 2)$. □

Proof of Theorem 1.3. When $k = 2$, for convenience, put $p^a = y, q^b = x$ for $(i, j) = (2, 3), k = 2$ and $p^a = x, q^b = y$ for $(i, j) = (3, 2), k = 2$ in (1.2), respectively. Let

$$x = \frac{X + 12}{12}, \ y = \frac{Y + 36}{72},$$

the converse transformation is

$$X = 12x - 12, \ Y = 36(2y - 1).$$

Then, we have

$$Y^2 = X^3 - 144X + 22032, \ Y^2 = X^3 - 144X - 19440,$$

respectively.

Using **Magma**, we get all the integral points on the above two elliptic curves. For $Y^2 = X^3 - 144X + 22032$, the points $(X, Y) = (36, 252), (24, 180)$ lead to the solutions $(p, q, a, b) = (2, 2, 2, 2), (3, 3, 1, 1)$ of (1.2). And $Y^2 = X^3 - 144X - 19440$ has no integral points, and hence (1.2) has no solution for $(i, j) = (3, 2)$.

Let $p^a = y, q^b = x$ for $(i, j) = (2, 4), k = 2$ in (1.2), let

$$x = X, \ y = \frac{Y + 3}{6},$$

we have

$$Y^2 = 3X^4 - 18X^3 + 33X^2 - 18X + 153.$$

Using **Magma**, we get all the integral points on this elliptic curve, which are

$$(X, \pm Y) = (-1, 4; 15).$$

We find that $(X, Y) = (4, 15)$ leads to the unique solutions $(p, q, a, b) = (3, 2, 1, 2)$ of (1.2).

Let $p^a = x, q^b = y$ for $(i, j) = (4, 2), k = 2$ in (1.2), let

$$x = X, \ y = \frac{Y + 3}{6},$$

we have

$$Y^2 = 3X^4 - 18X^3 + 33X^2 - 18X - 135.$$

Using **Magma**, we get all the integral points on this elliptic curve, which are

$$(X, \pm Y) = (-2, 5; 15),$$

by some calculations, we find that $(X, Y) = (5, 15)$ lead to the unique solution $(p, q, a, b) = (5, 3, 1, 1)$ of (1.2). □

In Figure 1, we display the graph of the quartic curve

$$Y^2 = 3X^4 - 18X^3 + 33X^2 - 18X - 135.$$

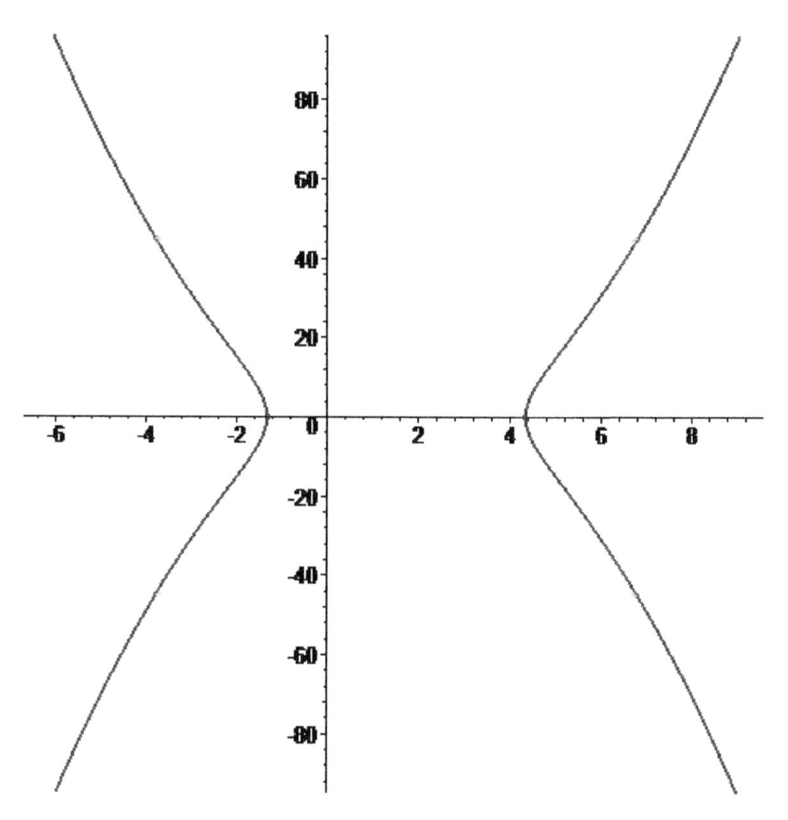

Fig. 1. $Y^2 = 3X^4 - 18X^3 + 33X^2 - 18X - 135$

4. Two conjectures related with Catalan equation

In [1], the first author raised a new variant of the Hilbert-Waring problem: to express a positive integer n as the sum of s positive integers whose product is a k-th power, i.e.,

$$n = x_1 + x_2 + \cdots + x_s$$

such that

$$x_1 x_2 \cdots x_s = x^k,$$

for $n, x_i, x, k \in \mathbb{Z}^+$, which may be regarded as a generalization of Waring's problem:

$$n = x_1^k + x_2^k + \cdots + x_s^k.$$

Now we expand this idea to Catalan's equation and consider

$$\begin{cases} A - B = 1, \\ AB \text{ is square-full}, \end{cases} \tag{4.1}$$

where A, B are positive integers.

By using the method of solving Pell's equation, it is easy to show that there are infinitely many solutions of (4.1), the least three being $(8, 9), (288, 289), (675, 676)$.

However, after calculations with computer, we raise the following conjectures (we have checked up to $B < A < 10^6$)

Conjecture 4.1. *Let $r \geq 0$ be an integer. Then the Diophantine equation*

$$\begin{cases} A - B = 2^r, \\ AB \textit{ is cube-full}, \end{cases}$$

has no solution for $r = 0$, and has a unique solution $(A, B) = (2^{r+1}, 2^r)$ for $r \geq 1$.

Moreover, we have

Conjecture 4.2. *Let $r \geq 1$ be an integer. Then there are infinitely many primes p such that the Diophantine equation*

$$\begin{cases} A - B = p^r, \\ AB \textit{ is cube-full}, \end{cases} \tag{4.2}$$

has no solution. Moreover, the least such a prime is 29.

It is easy to verify that for every integer $2 \leq n \leq 28$, (4.2) has solutions. However, we even do not know if there is a solution for infinitely many primes p.

Acknowledgements

The authors wish to thank Dr. Deyi Chen and Mr. Tanyue Gao for their kind help in calculation, especially in verifying Conjectures 2.1, 4.1 and 4.2. Project supported by the Natural Science Foundation of China (No. 11351002) and the Natural Science Foundation of Zhejiang Province (No. LQ13A010012).

References

1. T. Cai and D. Chen, A new variant of the Hilbert-Waring problem, Math. Comp. **82** (2013), no. 284, 2333–2341.
2. T. Cai, The Book of Numbers, High Education Press, Beijing, 2014.
3. R. K. Guy, Unsolved Problems in Number Theory, 3rd edition, Springer, 2004.
4. P. Mihăilescu, Primary cyclotomic units and a proof of Catalan's conjecture, J. Reine Angew. Math. **572** (2004), 167–195.

STATISTICAL DISTRIBUTION OF ROOTS OF A POLYNOMIAL MODULO PRIME POWERS

YOSHIYUKI KITAOKA

Department of Mathematics, Meijo University,
Tenpaku, Nagoya, 468-8502, Japan
E-mail: kitaoka@meijo-u.ac.jp

Let

$$f(x) = x^n + a_{n-1}x^{n-1} + \cdots + a_1 x + a_0 \tag{0.1}$$

be a polynomial with integer coefficients. Throughout this paper, unless otherwise specified, $f(x)$ is given in this form once and for all and the letter p denotes a prime number. It is well-known as Chebotarev's density theorem that the density of prime numbers p for which $f(x)$ is completely spitting over the p-adic number field $\mathbb{Q}_p$ is $1/[\mathbb{Q}(f):\mathbb{Q}]$, where $\mathbb{Q}(f)$ is a Galois extension of the rational number field $\mathbb{Q}$ generated by all roots of $f(x)$.

In this paper, we report several observations on distribution of roots of $f(x) \equiv 0 \bmod p^m$ when either p or m is fixed and the other tends to infinity.

Let us define fundamental notation. First,

$$\begin{aligned} Spl(f) &:= \{p \mid f(x) \text{ is completely splitting over } \mathbb{Q}_p\}, \\ Spl_X(f) &:= \{p \in Spl(f) \mid p \le X\}. \end{aligned}$$

In previous papers we used the term completely decomposable; however, in this paper to avoid confusion, we use the term 'completely splitting' and 'decomposable' is kept for a decomposable decomposition, see Footnote * below.

It is known that the difference between the set $Spl(f)$ and the set $\{p \mid f(x) \bmod p \text{ is completely splitting}\}$ is a finite set. For $p \in Spl(f)$, we decompose $f(x)$ as

$$f(x) \equiv \prod_{i=1}^{n} (x - r_{i,p^m}) \bmod p^m, \tag{0.2}$$

where we assume that rational integers r_{i,p^m} satisfy

$$0 \leq r_{1,p^m} \leq \cdots \leq r_{n,p^m} \leq p^m - 1. \tag{0.3}$$

At first, the introduction of this global order in the local situation may be unacceptable, but experiments suggest that it is, although no theoretical reason is found yet.

Secondly, it is easily seen that

$$a_{n-1} + \sum_i r_{i,p^m} \equiv 0 \bmod p^m, \tag{0.4}$$

whence there is an integer k_{p^m} such that

$$k_{p^m} p^m := a_{n-1} + \sum_i r_{i,p^m}. \tag{0.5}$$

This $k_{p^m} = k_{p^m}(f)$ and generalization thereof plays an important role in this paper. The following proposition is basic.

Prop 0.1. If $f(x)$ does not have a linear factor, then

$$1 \leq k_{p^m} \leq n - 1$$

for sufficiently large p^m.

Proofs of propositions are given in the last section.

If there is any relation among roots r_{i,p^m} besides the relation (0.4), the situation becomes complicated for experimental study of distribution of roots r_{i,p^m} and we exclude such exceptional cases. If $f(x)$ is reducible, that is $f(x) = g(x)h(x)$ for monic polynomials g, h, then relations corresponding to (0.4) for roots of $g(x), h(x)$ arise, and

$$k_{p^m}(f) = k_{p^m}(g) + k_{p^m}(h) \quad (p \in Spl(f) = Spl(g) \cap Spl(h)).$$

We exclude this case, and the next case is of type $f(x) = g(h(x))$ as follows:

Prop 0.2. Let $g(x), h(x) = x^v + h_{v-1}x^{v-1} + \ldots$ ($v := \deg h > 1$) be monic polynomials with integer coefficients, and suppose that $f(x) = g(h(x))$ has n distinct roots in $\mathbb{Q}_p$ for $p \in Spl(f)$. We re-label all roots r_{i,p^m} in (0.2) as $r_{i,j}$ ($1 \leq i \leq \deg g, 1 \leq j \leq v$) as follows:

$$g(x) \equiv \prod_{i=1}^{\deg g} (x - s_i) \bmod p^m, \quad h(x) - s_i \equiv \prod_{j=1}^{v} (x - r_{i,j}) \bmod p^m,$$

$$0 \leq {}^\exists s_i, r_{i,j} < p^m \ (1 \leq i \leq \deg g, 1 \leq j \leq v).$$

Then we have, for some integers k_{p^m} and k_l

$$a_{n-1} + \sum_{i,j} r_{i,j} = k_{p^m} p^m, \quad h_{v-1} + \sum_j r_{l,j} = k_l\, p^m \quad (1 \le {}^\forall l \le \deg g).$$

If f does not have a linear factor and p^m is sufficiently large, then we have

$$1 \le k_{p^m} \le n-1,\ 1 \le k_l \le v-1 \quad (1 \le {}^\forall l \le \deg g).$$

If $f(x) = g(h(x))$, $1 < v = \deg h < n = \deg f$, we call f *decomposable**, otherwise *indecomposable*. We refer to $f(x) = g(h(x))$ as a decomposable decomposition.

The two cases above were discussed in previous papers and the author assumed that they were the only exceptional polynomials, but polynomials given after Conjecture 1 below seem a part of other series of exceptional polynomials. The author does not know what they are, yet.

In [3]-[8], we made several observations in the case of $m = 1$. For convenience of readers, we summarize them with generalizations in §1 and §2.1.

1. Uniform distribution

In this section, a weaker condition $0 \le r_{1,p^m}, \ldots, r_{n,p^m} \le p^m - 1$ than (0.3) is sufficient and m is fixed. In previous papers, we conjectured the following expectation for irreducible and indecomposable polynomials f. Although there are additional exceptions as mentioned above, computer experiments support the truth of the expectation.

Expectation 1. Suppose f is irreducible and indecomposable. Put, for a natural number k with $1 \le k \le n-1$

$$Pr_X(f)[k] := \frac{\#\{p \in Spl_X(f) \mid k_{p^m} = k\}}{\#Spl_X(f)}.$$

Then up to unspecified exceptions, the limit $Pr(f)[k] = \lim_{X\to\infty} Pr_X(f)[k]$ exists and

$$Pr(f)[k] = E_n(k) := A(n-1,k)/(n-1)!.$$

Here $A(n,k)$ $(1 \le k \le n)$ are Eulerian numbers defined by

$$A(1,1) = 1,\ A(n,k) = (n-k+1)A(n-1,k-1) + kA(n-1,k)$$

*The author learned from Professor Schinzel that what we called a reduced polynomial in previous papers is called a decomposable polynomial and there is an introduction to Ritt's theory about decomposable polynomials in his book [9].

and $E_n(k)$ is the volume of the set $\{x \in [0,1)^{n-1} \mid \lceil x_1 + \cdots + x_{n-1} \rceil = k\}$. Here, $\lceil x \rceil$ denotes the least integer greater than or equal to x. (Data for Expectation 1 in the case of $m = 1$ are given in [4].) The values of Eulerian numbers are

$n \setminus k$	1	2	3	4	5	6
1	1					
2	1	1				
3	1	4	1			
4	1	11	11	1		
5	1	26	66	26	1	
6	1	57	302	302	57	1

In what follows we shall introduce other densities, in the same spirit as in Expectation 1, $Pr(\cdots)$ with suffix X or without, the latter meaning the limit as $X \to \infty$ if it exists and most of them are with respect to $\#Spl_X(f)$.

Let us give a few comments on Expectation 1 and uniform distribution of roots modulo p:

(i) It is known that $r_{i,p}/p$ is uniformly distributed in $[0,1)$ if $n = 2$ ([1], [10]).

(ii) We see easily that n-tuples $(r_{1,p^m}/p^m, \ldots, r_{n,p^m}/p^m) \in [0,1)^n$ are not uniformly distributed, because $\sum_{i=1}^{n} r_{i,p^m}/p^m - k_{p^m} = -a_{n-1}/p^m \to 0$ when $p^m \to \infty$.

(iii) The several observations suggest that *all* $n!$ *points* $(r_{\sigma(1),p^m}/p^m, \ldots, r_{\sigma(n-1),p^m}/p^m)$ $(^\forall \sigma \in S_n)$ *are uniformly distributed in* $[0,1)^{n-1}$ *for many irreducible and indecomposable polynomials* f. (Once we have given the ordering of roots by (0.3), we are also concerned with the distribution of roots under a fixed permutation σ).

(iv) The uniformity above implies Expectation 1: If f does not have a linear factor, then the equation $k_{p^m} = \lceil r_{\sigma(1),p^m}/p^m + \cdots + r_{\sigma(n-1),p^m}/p^m \rceil$ holds except finitely many primes p, whence we see that if the uniformity above is true for a polynomial f, then $Pr(f)[k]$ is equal to the volume $E_n(k)$ of $\{x \in [0,1)^{n-1} \mid \lceil x_1 + \cdots + x_{n-1} \rceil = k\}$ by the definition of the uniformity, that is Expectation 1 is true.

We can say a bit in the decomposable case. If $f(x) = g(h(x))$ holds for a monic polynomial $h(x)$ with $h(0) = 0$, $\deg h$ determines the polynomial h itself ([6]). Suppose that a decomposable decomposition is unique, i.e. $\deg h$ is unique, and that using notations in Proposition 0.2, all arrangements of $(\deg h - 1)$-tuples of roots of $h(x) - s_i \equiv 0 \bmod p^m$ $(i = 1, \ldots, \deg g)$ are uniformly distributed in $[0,1)^{(\deg h - 1)\deg g}$; then the density is given by the convolution product $E_v^{\deg g}$ (cf. (4) after Remark 1.2 in [5]), i.e.

$$Pr(f)[k] = \sum_{i_1+\cdots+i_{\deg g}=k} E_v(i_1)\cdots E_v(i_{\deg g}). \tag{1.1}$$

This also matches with experiments well ([5]).

We have no perspective in case that there are plural decomposable decompositions. Let us give examples.

(i) Let $f(x) = h_1(h_1+c)^3 + d = h_2^3 + d$ ($h_1 = x^3, h_2 = x^4 + cx$), which has two decomposable decompositions. Conjectural densities are given by $(Pr(f)[4], \ldots, Pr(f)[8]) = (1/15, 7/30, 2/5, 7/30, 1/15)$. $Pr(f)[k] = 0$ for other integers k ([5]). Since there are relations corresponding to h_1, h_2, it is not strange that the density is different from (1.1), but we cannot elucidate the meaning of values.

(ii) Another example is $f(x) = (x^3)^5 + 2 = (x^5)^3 + 2$, for which the density is given by (1.1) with $\deg h = 3, \deg g = 5$ as if relations from $f(x) = (x^5)^3 + 2$ do not exist ([5]). This observation is based only on the distribution of k_p (there is no observation on the distribution of roots themselves).

(iii) We have the following

Prop 1.1. If a polynomial $f(x) = g(h(x))$ is irreducible and $\deg h = 2$, then we have $k_{p^m} = n/2$ for a sufficiently large p^m, that is the density is given by (1.1), even if there are other decomposable decompositions.

It is a problem for the time being to find irreducible and indecomposable polynomials for which Expectation 1 does not hold.

Remark 1.1. There are some data for reducible polynomials in [5].

2. With congruence relations

In this section we recall the first motivation for our research which is concerned with the distribution of roots with a congruence condition, and then go on to presenting a new problem modulo prime powers with a congruence condition. Denote the maximal abelian extension over the rational number field $\mathbb{Q}$ by $\mathbb{Q}^{ab}$, which is generated by all kth roots of unity ($k = 2, 3, \ldots$) by the Kronecker-Weber theorem.

Let F be an abelian extension of $\mathbb{Q}$ and take a natural number c such that $F \subset \mathbb{Q}(\zeta_c)$, where ζ_c is a primitive cth root of unity. For an integer a relatively prime to c, $[[a]]$ denotes the automorphism of F induced by $\zeta_c \to \zeta_c^a$.

2.1. *From [8]*

First, let us quote a result on periodic expansions of rationals ([7]), which is a starting point of our series of experiments.

Theorem 2.1. *Let a, b be natural numbers and suppose that $(10a, b) = 1$, $a < b$, and let $e = nk$ be the least length of periods of the decimal expansion of a/b with integers $n(> 1), k$. Putting*

$$a/b = 0.\dot{c}_1 \dots \dot{c}_e \quad (0 \leq c_i \leq 9),$$
$$L = (10^k - 1, b),\ b = BL,$$

we have

$$c_1 \dots c_k + c_{k+1} \dots c_{2k} + \cdots + c_{(n-1)k+1} \dots c_{nk} = \kappa \cdot (10^k - 1)/L,$$

where κ is the natural number defined by

$$\kappa = \sum_{i=0}^{n-1} r_i/B$$

where

$$r_i \equiv 10^{ki} a \bmod b \ (0 \leq r_i < b).$$

If $a = 1$ and $n = 2$, then κ is determined by

$$B\kappa \equiv 2 \bmod L, 1 \leq \kappa \leq L.$$

Remark 2.1. e is equal to the order of 10 mod b. Let E be the order of 10 mod L and decompose $k = KE$, noting $10^k \equiv 1 \bmod L$. Then we see that

$$(10^k - 1)/L = ((10^E)^{K-1} + (10^E)^{K-2} + \cdots + 1)(10^E - 1)/L,$$

which is a K-times iteration of the least period $C_1 \dots C_E \, (= (10^E - 1)/L)$ of $1/L$. If $L = 1$, then $(10^k - 1)/L = 9 \dots 9$.
κ is random in general, and the distribution of κ was studied in [3].
Putting $f(x) = x^n - a^n$, then the $r_i's$ in Theorem 2.1 satisfy $f(r_i) \, (\equiv 0 \bmod b) \equiv 0 \bmod B$ and $r_i \equiv a \bmod L$. We generalize this situation to a general polynomial, assuming $B = p$ in the rest of this subsection.

We consider the following density for an irreducible polynomial f: Let L be a natural number greater than 1, and for a prime $p \in Spl(f)$ with $p \nmid L$, we take roots $r_{i,p}$ of $f(x) \equiv 0 \bmod p$ so that

$$\begin{cases} f(r_{i,p}) \equiv 0 \bmod p, \\ r_{i,p} \equiv 0 \bmod L, \qquad (i = 1, \dots, n) \\ 0 \leq r_{i,p} \leq pL - 1, \end{cases} \tag{2.1}$$

by Chinese remainder theorem. The ordering (0.3) is replaced by the third condition above. We define an integer k_p by $a_{n-1} + \sum r_{i,\,p} = k_p p$ similarly to (0.5), and put

$$Pr_X(f, L)[k] := \frac{\#\{p \in Spl_X(f) \mid k_p = k\}}{\#Spl_X(f)}.$$

If p is large, then we have $1 \leq k_p \leq nL - 1$ similarly to Proposition 0.1. Let us give observations on $Pr(f, L)[k] := \lim_{X\to\infty} Pr_X(f, L)[k]$. First, we introduce a function

$$F_n(x) := \frac{1}{(n-1)!} \sum_{0\leq i\leq x} (-1)^i \binom{n}{i} (x-i)^{n-1},$$

where i runs over integers. It is known ([2]) that this is the volume of

$$\{(x_1, \ldots, x_{n-1}) \in [0,1)^{n-1} \mid x - 1 < \sum_{i=1}^{n-1} x_i \leq x\},$$

whence

$$F_n(k) = E_n(k) \quad (1 \leq k \leq n-1, k \in \mathbb{Z})$$

with $E_n(k)$ defined above. Putting $N = (a_{n-1}, L)$, $T = L/N$, we see that the condition $Pr_X(f, L)[j] \neq 0$ for a large number X implies (i) $j < nL$, (ii) $(j, L) = N$, and (iii) that a_{n-1}/N and j/N induce the same automorphism $[[a_{n-1}/N]] = [[j/N]]$ on $\mathbb{Q}(f) \cap \mathbb{Q}(\zeta_T)$. The following expectation is plausible for many irreducible and indecomposable polynomials.

Expectation 2. For $1 \leq j < nT$,

$$Pr(f, L)[jN] = \begin{cases} \dfrac{[\mathbb{Q}(f) \cap \mathbb{Q}(\zeta_T) : \mathbb{Q}]}{\varphi(T)} F_n(j/T) & \text{if } (j, T) = 1 \text{ and } [[j]] = [[a_{n-1}/N]], \\ 0 & \text{otherwise,} \end{cases}$$

with $\varphi(T)$ indicating the Euler function.

For an irreducible and indecomposable polynomial, the above expectation is the main conjecture of [8], but unfortunately it is not true for polynomials of degree 6 given after Conjecture 1.

We may consider (2.1) in the wider setting with p^m instead of p.

2.2. *Prime power modulus*

From here, let us give a new kind of observations. The study of the distribution of roots r_{i,p^m} subject to the following condition is more interesting, but more difficult than that subject to (2.1): I.e., given integers R_i with $0 \le R_i \le L-1$, let us consider the following condition besides (0.2), (0.3)

$$r_{i,\,p^m} \equiv R_i \bmod L \quad (i = 1, \ldots, n). \tag{2.2}$$

We are concerned with the density

$$Pr(f, m, L, \{R_i\}) := \lim_{X \to \infty} Pr_X(f, m, L, \{R_i\}), \tag{2.3}$$

where

$$Pr_X(f, m, L, \{R_i\}) := \frac{\#\{p \in Spl_X(f) \mid (0.2), (0.3), (2.2)\}}{\#Spl_X(f)}.$$

As a necessary condition for the existence of roots satisfying (2.2), we find

Prop 2.1. Assume that $f(x)$ is irreducible and let m, L be natural numbers, and q an integer relatively prime to L. Put

$$R_f := a_{n-1} + \sum_{i=1}^{n} R_i, \quad d := (R_f, L). \tag{2.4}$$

We assume the decomposition (0.2) for a prime $p \in Spl(f)$ with $p \equiv q$ mod L and a weaker condition than (2.2):

$$\sum_{i=1}^{n} r_{i,p^m} \equiv \sum_{i=1}^{n} R_i \bmod L.$$

If p^m is sufficiently large, then $[[q]] = [[1]]$ on $\mathbb{Q}(f) \cap \mathbb{Q}(\zeta_{L/d})$ and there is an integer K which satisfies the conditions

$$\begin{cases} R_f \equiv Kq^m \bmod L, \\ 1 \le K \le n-1, (K, L) = d. \end{cases} \tag{2.5}$$

Remark 2.2. The condition "$[[q]] = [[1]]$ on $\mathbb{Q}(f) \cap \mathbb{Q}(\zeta_{L/d})$" is trivially satisfied if $\mathbb{Q}(f) \cap \mathbb{Q}(\zeta_{L/d}) = \mathbb{Q}$.

The case of $\deg f = 1$ is as follows:

Prop 2.2. Let $f(x) = x + a$ ($a \in \mathbb{Z}$), and $0 \le R \le L-1$. In case of $a > 0$, the density $Pr(f, m, L, R)$ is non-zero if and only if $R + a \equiv l^m \bmod L$ for some integer l with $(l, L) = 1$, and then the density is equal to $\#\{q \bmod L \mid q^m \equiv 1 \bmod L\}/\varphi(L)$. In case of $a \le 0$, the density is 1 if $R + a \equiv 0 \bmod L$, and is 0 otherwise.

The proof shows that this is nothing but Dirichlet's prime number theorem for an arithmetic progression.

Next, let us define a naive conjectural density $pr(f, m, L, \{R_i\})$.

Definition 2.1. Suppose that $m, L, R_1, \ldots, R_n$ are integers with $m \ge 1, L \ge 2, 0 \le R_i \le L - 1$ and $f(x)$ is a monic polynomial of degree $n \ge 2$. With (2.4) we put

$$pr(f, m, L, \{R_i\}) := \sum_{K, q} E_n(K) \cdot \frac{1}{L^{n-1}} \cdot \frac{1}{[\mathbb{Q}(\zeta_L) : \mathbb{Q}(f) \cap \mathbb{Q}(\zeta_{L/d})]}, \tag{2.6}$$

where $q \in (\mathbb{Z}/L\mathbb{Z})^\times$ and K satisfy the conditions in Proposition 2.1, i.e.

$$\begin{cases} [[q]] = [[1]] \text{ on } \mathbb{Q}(f) \cap \mathbb{Q}(\zeta_{L/d}), & (2.5.1) \\ R_f \equiv K q^m \bmod L, & (2.5.2) \\ 1 \le K \le n - 1, (K, L) = d. & (2.5.3) \end{cases}$$

Remark 2.3. The above density pr depends only on the field $\mathbb{Q}(f)$ and the sum R_f mod L, not on the polynomial f and $R_1, \ldots, R_n$ themselves.

Remark 2.4. Suppose that $n = 2$; then $pr(f, m, L, \{R_i\}) = 0$ if $(R_f, L) \ne 1$. If $(R_f, L) = 1$,

$$pr(f, m, L, \{R_i\}) \\ = \frac{1}{\varphi(L^2)} \begin{cases} \#\{q \bmod L \mid R_f \equiv q^m \bmod L\} & \text{if } D \nmid L, \\ 2\#\{q \bmod L \mid R_f \equiv q^m \bmod L, \chi_D(q) = 1\} & \text{if } D \mid L, \end{cases}$$

where D is the discriminant of the quadratic field $\mathbb{Q}(f)$ and χ_D is the Kronecker symbol $(\frac{D}{*})$.

Remark 2.5. The motivation for the formulation of density (2.6) is as follows. The number of $q \in (\mathbb{Z}/L\mathbb{Z})^\times$, which satisfies $[[q]] = [[1]]$ on $\mathbb{Q}(f) \cap \mathbb{Q}(\zeta_{L/d})$ is $[\mathbb{Q}(\zeta_L) : \mathbb{Q}(f) \cap \mathbb{Q}(\zeta_{L/d})]$, and since an n-tuple $(R_1, \ldots, R_n)$ is restricted by the condition $a_{n-1} + R_1 + \cdots + R_n \equiv R_f \bmod L$ for a given R_f, the density of n-tuples $(R_1, \ldots, R_n)$ is $1/L^{n-1}$, and lastly the density of primes for K is $E_n(K)$.
If $(R_f, L) \ge n$, then $pr(f, m, L, \{R_i\}) = 0$ by $d = (K, L) \le n - 1$.

The following proposition shows that the sum of above conjectural densities over the set $\{(R_1, \ldots, R_n) \mid 0 \le R_1, \ldots, R_n \le L - 1\}$ is surely 1.

Prop 2.3. We have $\sum_{0 \le R_1,\dots,R_n \le L-1} pr(f,m,L,\{R_i\}) = 1$.

We note that $(n-1)! \cdot L^{n-1} \cdot \varphi(L) \cdot pr(f,m,L,\{R_i\})$ is an integer, and hence if $Pr(f,m,L,\{R_i\}) = pr(f,m,L,\{R_i\})$ holds, then for a large number X, the nearest integer to $(n-1)! \cdot L^{n-1} \cdot \varphi(L) \cdot Pr_X(f,m,L,\{R_i\})$ should be equal to $(n-1)! \cdot L^{n-1} \cdot \varphi(L) \cdot pr(f,m,L,\{R_i\})$. We have checked the existence of such an X in the following cases:

(1) Irreducible polynomials $f(x) = x^2 + a_1x + a_0$ with $0 \le |a_1|, |a_0| \le 50$, $1 \le m \le 5$ and $2 \le L \le 30$.
(2) Irreducible polynomials $f(x) = x^3 + a_2x^2 + a_1x + a_0$ with $0 \le a_2, |a_1|, |a_0| \le 10$, $1 \le m \le 2, 2 \le L \le 15$.
(3) With $1 \le m \le 5$, $2 \le L \le 10$, irreducible and indecomposable polynomials $x^4+x^3+x^2+x+1, x^4+2x^2+16x+17, x^4+x^3+x^2-x+1, x^4+2x^3+2x^2+2, x^4+x^3+x^2+x-1$ whose Galois group is cyclic, biquadratic, dihedral, alternating, and symmetric, respectively, in this order and polynomials $x^4-x^3+2x^2+x+1$, $x^4-x^3+5x^2+2x+4$, $x^4-x^3-x^2-2x+4$ corresponding to biquadratic fields.
(4) Polynomial $f(x) = x^5 - 10x^3 + 5x^2 + 10x + 1$, which defines a subfield of $\mathbb{Q}(\zeta_{25})$, with $m = 1, 2$ and $2 \le L \le 5$.

Conjecture 1. *If a polynomial $f(x)$ is irreducible and indecomposable,*

$$Pr(f,m,L,\{R_i\}) = pr(f,m,L,\{R_i\}) \text{ if } \deg f \le 5.$$

Although data may be insufficient in the case of degree 4, 5, the author found, by a time-saving rough checking method, no candidates of polynomials with $Pr \ne pr$.

If $\deg f = 6$, then the conjecture is false for the irreducible and indecomposable polynomial $f_1 = x^6 - 2x^5 + 18x^4 - 22x^3 + 163x^2 - 116x + 631$, but $Pr(f_2,m,L,\{R_i\}) = pr(f_2,m,L,\{R_i\})$ holds for $f_2 = x^6 + x^5 + x^4 + x^3 + x^2 + x + 1$ (the cases $m = 1, 2$, and $L = 2, 3, 4$ are checked). Both polynomials define the same field $\mathbb{Q}(\zeta_7)$.
Let us give data: For f_2 with $m = 1$, $L = 2$:

$$pr = \begin{cases} \dfrac{26}{5!} \cdot \dfrac{1}{2^5} + \dfrac{26}{5!} \cdot \dfrac{1}{2^5} = \dfrac{13}{960} & \text{if } R_f \equiv 0 \bmod L, \\ \dfrac{1}{5!} \cdot \dfrac{1}{2^5} + \dfrac{66}{5!} \cdot \dfrac{1}{2^5} + \dfrac{1}{5!} \cdot \dfrac{1}{2^5} = \dfrac{17}{960} & \text{if } R_f \equiv 1 \bmod L. \end{cases}$$

For f_1 with $m = 1$, $L = 2$: The table is given below whose first line is to mean: for $(R_1, \dots, R_6) = (0,1,1,1,1,0), (1,0,0,0,0,1)$, their density is expected to be $6/d$, where $d = 2160$.

$6/d : (0,1,1,1,1,0), (1,0,0,0,0,1),$
$12/d : (0,1,0,0,0,1), (0,1,1,1,0,1), (1,0,0,0,1,0), (1,0,1,1,1,0),$
$16/d : (0,0,1,0,0,1), (0,1,1,0,1,1), (1,0,0,1,0,0), (1,1,0,1,1,0),$
$20/d : (0,0,0,1,1,0), (0,1,1,0,0,0), (1,0,0,1,1,1), (1,1,1,0,0,1),$
$22/d : (0,0,0,1,0,1), (0,1,0,1,1,1), (1,0,1,0,0,0), (1,1,1,0,1,0),$
$23/d : (0,0,0,1,0,0), (0,0,1,0,0,0), (1,1,0,1,1,1), (1,1,1,0,1,1),$
$25/d : (0,0,1,0,1,0), (0,0,1,1,0,1), (0,1,0,0,1,1), (0,1,0,1,0,0),$
$(1,0,1,0,1,1), (1,0,1,1,0,0), (1,1,0,0,1,0), (1,1,0,1,0,1),$
$26/d : (0,0,0,0,1,0), (0,1,0,0,0,0), (1,0,1,1,1,1), (1,1,1,1,0,1),$
$33/d : (0,0,1,0,1,1), (1,1,0,1,0,0),$
$35/d : (0,1,0,0,1,0), (1,0,1,1,0,1),$
$39/d : (0,0,1,1,1,0), (0,1,1,1,0,0), (1,0,0,0,1,1), (1,1,0,0,0,1),$
$40/d : (0,0,0,0,1,1), (0,0,1,1,1,1), (1,1,0,0,0,0), (1,1,1,1,0,0),$
$41/d : (0,0,0,0,0,1), (0,1,1,1,1,1), (1,0,0,0,0,0), (1,1,1,1,1,0),$
$48/d : (0,0,0,1,1,1), (1,1,1,0,0,0),$
$49/d : (0,0,1,1,0,0), (1,1,0,0,1,1),$
$58/d : (0,1,0,1,1,0), (0,1,1,0,1,0), (1,0,0,1,0,1), (1,0,1,0,0,1),$
$60/d : (0,0,0,0,0,0), (1,1,1,1,1,1),$
$69/d : (0,1,0,1,0,1), (1,0,1,0,1,0),$
$86/d : (0,1,1,0,0,1), (1,0,0,1,1,0)$

Densities for $(R_1,\ldots,R_6), (1-R_1,\ldots,1-R_6), (1-R_6,\ldots,1-R_1)$ are the same.

Let us give a density table for Pr in (2.3) in contrast to pr above, although data are insufficient and there may be other possibilities.

Suppose $m = 1$, $L = 2$. A 6-tuple $(R_1,\ldots,R_6)$ with $0 \le R_i \le 1$ corresponds to an integer r with $1 \le r \le 2^6$ by

$$r = 1 + \sum R_i 2^{i-1}. \tag{2.7}$$

Let us explain the table below, referring to type (1).

(i) The rth coordinate of Pr is the density for $(R_1,\ldots,R_6)$ under the above correspondence (2.7). According as $Pr_i = Pr_{2^6+1-i}$ holds for $^\forall i$ or not, the proviso "symmetric" or "non-symmetric" is added. For $r = 1$, $36/2304$ is the density for $R = (0,\ldots,0)$ and Pr is non-symmetric.

(ii) "Values" mean the arrangement of numerators of densities in ascending order.
(iii) If the sum of the coordinates in opposite positions in the vector "values" in (ii) is the same, it is stated as "symmetric sum value". I.e. if l is the length of the vector, and the sum of ith and $(l-i+1)$th coordinates of $values$ are the same, then it gives symmetric sum value. In the fist line, it is 72.
(iv) If the densities for $(R_1, \ldots, R_6)$ and $(L-1-R_6, \ldots, L-1-R_1)$ are the same, the proviso $(R_1, \ldots) = (L-1-R_6, \ldots)$ is added.
(v) Polynomials with those densities are given.

(1):
(i) $Pr = (36, 4, 15, 43, 43, 42, 23, 62, 29, 30, 35, 48, 36, 38, 49, 43, 57, 29, 36, 38, 37, 40, 29, 42, 49, 54, 43, 30, 23, 29, 36, 4, 68, 36, 43, 49, 42, 29, 18, 23, 30, 43, 32, 35, 34, 36, 43, 15, 29, 23, 34, 36, 24, 37, 42, 43, 10, 49, 30, 29, 29, 57, 68, 36)/2304$: non-symmetric.
(ii) $values = (4, 10, 15, 18, 23, 24, 29, 30, 32, 34, 35, 36, 37, 38, 40, 42, 43, 48, 49, 54, 57, 62, 68)$.
(iii) symmetric sum of values $= 72$.
(iv) $(R_1, \ldots) = (L-1-R_6, \ldots)$.
(v) $x^6 - 2x^5 - 2x^4 + 24x^3 + 86x^2 + 178x + 215, x^6 - 2x^5 - 14x^4 + 14x^2 - 2x - 1, x^6 - 2x^5 + 8x^4 + 11x^3 + 54x^2 + 63x + 81, x^6 - 2x^5 - 22x^4 + 67x^3 - 30x^2 - 33x + 11, x^6 - 2x^5 - 16x^4 + 49x^3 - 24x^2 - 15x - 1, x^6 - 2x^5 + 4x^4 - 18x^3 + 70x^2 - 128x + 109, x^6 - 2x^5 + 4x^4 - 9x^3 + 30x^2 - 9x + 9, x^6 - 2x^5 - 2x^3 + 30x^2 - 52x + 29, x^6 - 2x^5 - 10x^4 + 16x^3 + 150x^2 + 222x + 131, x^6 - 2x^5 - 6x^4 + x^3 + 80x^2 + 21x + 9, x^6 - 2x^5 - x^4 + 10x^3 + 22x^2 - 8x + 16, x^6 - 2x^5 - 4x^4 + 12x^3 + 70x^2 + 124x + 83, x^6 - 2x^5 - 8x^4 + 9x^3 + 152x^2 - 527x + 527, x^6 - 2x^5 - 14x^4 + 19x^3 + 238x^2 - 773x + 747, x^6 - 2x^5 + 5x^4 + 4x^3 + 34x^2 + 16x + 16$.

(2):
(i)
$Pr = (36, 68, 57, 29, 29, 30, 49, 10, 43, 42, 37, 24, 36, 34, 23, 29, 15, 43, 36, 34, 35, 32, 43, 30, 23, 18, 29, 42, 49, 43, 36, 68, 4, 36, 29, 23, 30, 43, 54, 49, 42, 29, 40, 37, 38, 36, 29, 57, 43, 49, 38, 36, 48, 35, 30, 29, 62, 23, 42, 43, 43, 15, 4, 36)/2304$: non-symmetric.
(ii)
$values = (4, 10, 15, 18, 23, 24, 29, 30, 32, 34, 35, 36, 37, 38, 40, 42, 43, 48, 49, 54, 57, 62, 68)$.
(iii) symmetric sum of values $= 72$.
(iv) $(R_1, \ldots) = (L-1-R_6, \ldots)$.
(v) $x^6 - 2x^3 + 9x^2 + 6x + 2$.

(3):
(i) $Pr = (7,9,9,7,9,7,7,9,9,7,7,9,7,9,9,7,9,7,7,9,7,9,9,7,7,9,9,7,$ $9,7,7,9,9,7,7,9,7,9,9,7,7,9,9,7,9,7,7,9,7,9,9,7,9,7,7,9,9,7,7,9,7,9,9,$ $7)/512$: symmetric.
(iv) $(R_1,\dots) = (R_6,\dots), (R_1,\dots) = (L-1-R_1,\dots), (R_1,\dots) = (L-1-R_6,\dots)$.
(v) $x^6+4x^5+1, x^6+2x^5+2, x^6+4x^5+2, x^6+4x^5+x+2$.

(4):
(i) $Pr = (9,7,7,9,7,9,9,7,7,9,9,7,9,7,7,9,7,9,9,7,9,7,7,9,9,7,7,9,$ $7,9,9,7,7,9,9,7,9,7,7,9,9,7,7,9,7,9,9,7,9,7,7,9,7,9,9,7,7,9,9,7,9,7,7,$ $9)/512$: symmetric.
(iv) $(R_1,\dots) = (R_6,\dots), (R_1,\dots) = (L-1-R_1,\dots), (R_1,\dots) = (L-1-R_6,\dots)$.
(v) x^6+5x^5+1, x^6+5x^5+x+2.

(5):
(i) $Pr = (133,48,16,105,16,58,45,40,16,57,73,44,138,32,52,105,16,39,$ $100,32,73,81,76,58,45,102,76,57,52,39,48,48,48,48,39,52,57,76,102,$ $45,58,76,81,73,32,100,39,16,105,52,32,138,44,73,57,16,40,45,58,16,$ $105,16,48,133)/3840$: symmetric.
(ii)
$values = (16,32,39,40,44,45,48,52,57,58,73,76,81,100,102,105,133,$ $138)$.
(iv) $(R_1,\dots) = (R_6,\dots), (R_1,\dots) = (L-1-R_1,\dots), (R_1,\dots) = (L-1-R_6,\dots)$.
(v) $x^6-3x^5+6x^4+3x^3-9x^2-18x+36$.

Type (1) and type (2) resp. type (3) and type (4) seem to form a pair. Is there a type corresponding to (5) to be a pair? Data are insufficient to say anything general.
The author does not know how to distinguish polynomials whose densities are given by pr in Definition 2.1.

3. Other densities

In this section, we study two other densities for a monic polynomial $f(x) \in \mathbb{Z}[x]$.

3.1.

Numerical data suggest that the distribution of k_{p^m} $(m \to \infty)$ is identical with the distribution of k_{p^m} $(p \in Spl(f) \to \infty)$ in [5], that is

Conjecture 2. *For a monic polynomial f,*

$$\lim_{X\to\infty}\frac{\#\{p\in Spl_X(f)\mid k_{p^m}=k\}}{\#\{p\in Spl_X(f)\}}\qquad (m:\textit{fixed},\ p:\textit{run})$$
$$=\lim_{X\to\infty}\frac{\#\{m\le X\mid k_{p^m}=k\}}{\#\{m\le X\}}\qquad (m:\textit{run},\ p(\in Spl(f)):\textit{fixed})$$

for any integer k.

The above conjecture may suggest: Putting

$$\mathbb{N}_f := \{l\in\mathbb{N}\mid f(x)\equiv\prod_{i=1}^{n}(x-r_i)\bmod l\},$$

we had better to consider the density

$$\lim_{X\to\infty}\frac{\#\{l\in\mathbb{N}_f\mid l<X,\ k_l=k\}}{\#\{l\in\mathbb{N}_f\mid l<X\}},$$

where k_l is defined for l similarly to (0.5) instead of p^m.

3.2.

Let us consider one more density. Let $L\,(>1)$ be a natural number relatively prime to $q\,(\in Spl(f))$, and $0\le R_1,\dots,R_n\le L-1$ non-negative integers. For an integer c, we put

$$Pr_X(f,q,L,\{R_i\};c)$$
$$:=\frac{\#\left\{m\le X\ \middle|\ \begin{array}{l} m\equiv c\bmod\varphi(L),\\ f(x)\equiv\prod(x-r_i)\bmod q^m\ (0\le r_1\le\cdots\le r_n<q^m),\\ r_i\equiv R_i\bmod L\ (i=1,\dots,n)\end{array}\right\}}{\#\{m\le X\mid m\equiv c\bmod\varphi(L)\}}.$$

The condition $Pr_X>0$ implies that $R_f\equiv Kq^c\bmod L$ with $1\le{}^{\exists}K\le n-1$ by Proposition 2.1. Let us give the final expectation.

Expectation 3. Let a polynomial $f(x)=x^n+a_{n-1}x^{n-1}+\cdots\in\mathbb{Z}[x]$ be irreducible. Then the limit $\lim_{X\to\infty}Pr_X(f,q,L,\{R_i\};c)$ exists, and the non-vanishing of it is equivalent to $R_f\equiv kq^c\bmod L$ with $1\le{}^{\exists}k\le n-1$, and then

$$Pr(f,q,L,\{R_i\};c)(:=\lim_{X\to\infty}Pr_X(f,q,L,\{R_i\};c))=1/(n-1)L^{n-1}.\quad(3.1)$$

The sum of the right-hand in (3.1) under the condition $0 \le R_1, \dots, R_n \le L-1$ ($R_i \in \mathbb{Z}$) is surely 1.
The conjecture is supported by numerical data in the case of $f = x^2 + a_1x + a_0$ ($-5 \le a_1 \le 10, |a_0| \le 10$, $2 \le L \le 15, 0 \le c \le 7$) by checking that for $X = 10^5$ and several primes p, the nearest integer of the left-hand side multiplied by $(n-1)L^{n-1}$ $(= L)$ is 1. And it is checked for polynomials $f = x^3+2$, x^3+x^2-2x-1 and $0 \le c \le 7, 2 \le L \le 15$ with $X = 5 \cdot 10^5$, too. Even in this case, it takes much time to make data, and for $n = 4$, $X = 10^6$ is small to get accurate data. There is no perspective when a polynomial is decomposable.

Remark 3.1. Let us consider local solutions r_i $(1 \le i \le n)$ of $f(x)$ in $\mathbb{Q}_p$, and write as $r_i = r_{i,0}+r_{i,1}p+\dots$ $(0 \le r_{i,0}, r_{i,1}, \dots < p)$. The distribution of $\{r_{1,m}, \dots, r_{n,m}\}$ may be interesting, too. How about a polynomial in many variables?

4. Proofs

Proof of Proposition 0.1. We abbreviate $a_{n-1}, r_{i,p^m}, k_{p^m}$ to a, r_i, k, respectively; then it is obvious

$$a/p^m + \sum_{i=1}^{n} r_i/p^m = k,$$

and hence the inequality (0.3) implies easily $0 \le k \le n$ if p^m is sufficiently large. Let us show furthermore that $k \ne 0, n$. Suppose that there are infinitely many numbers p^m such that $k = 0$; then we have $0 \le r_1 \le \sum r_i = -a$. Therefore there is an integer R $(0 \le R \le -a)$ such that $r_1 = R$ for infinitely many p^m. It implies $f(R) \equiv 0 \bmod p^m$ for infinitely many p^m, that is $f(R) = 0$. It contradicts the assumption on $f(x)$. Next, suppose $k = n$ for infinitely many p^m; then the equation $a = \sum(p^m - r_i)$ implies $0 < p^m - r_1 < a$, and there is an integer R' with $0 < R' < a$ such that $p^m - r_1 = R'$ for infinitely many p^m, which implies $f(-R') \equiv f(r_1) \equiv 0 \bmod p^m$, and hence $f(-R') = 0$, which is a contradiction, too. □

Proof of Proposition 0.2. Decompose as $f(x) = g(h(x)) = \prod(x - \tilde{r}_i)$ for distinct p-adic integers $\tilde{r}_i$, and put $\{\tilde{s}_i \mid i = 1, \dots, \deg g\} = \{h(\tilde{r}_i) \mid i = 1, \dots, n\}$. Then we have $g(x) = \prod(x - \tilde{s}_i)$ and $f(x) = \prod(h(x) - \tilde{s}_i)$. By reduction modulo p^m, we complete the proof of the first part. The inequality $1 \le k_{p^m} \le n-1$ follows from Proposition 0.1, and $0 \le k_i \le v$

is proved similarly. But we need revision to the proof of $1 \le k_i \le v-1$, since a factor $h(x) - s_i$ depends on p^m. If $k_i = v$ holds, then we have $h_{v-1} = \sum_j (p^m - r_{i,j})$, hence $0 \le p^m - r_{i,1} \le h_{v-1}$. These inequalities for infinitely many numbers p^m imply the existence of an integer R such that $R = p^m - r_{i,1}$, from which $h(-R) - s_i \equiv 0 \bmod p^m$ follows, whence we get $f(-R) \equiv 0 \bmod p^m$. Thus we have a contradiction that an integer $-R$ is a root of $f(x)$. If $k_i = 0$ happens, then we have $0 \le r_{i,1} \le \sum_j r_{i,j} = -h_{v-1}$. Similarly, infinitely many such inequalities imply an existence of an integer root of $f(x)$, which is a contradiction again. Thus, $1 \le k_i \le v-1$. □

Proof of Proposition 1.1. The proof is similar to that of Theorem 1.4 in [4]. We may suppose that $h(x) = (x+a)^2$ or $h(x) = (x+a)(x+a+1)$ for some integer a. Put $f(x) \equiv \prod_{i=1}^{n/2}(h(x) - s_i) \bmod p^m$ similarly to Proposition 0.2. First we assume that $h(x) = (x+a)^2 = x^2 + 2ax + a^2$, whence $f(x) = x^n + anx^{n-1} + \dots$. We take and fix i $(1 \le i \le n/2)$. Since $h(x-a) - s_i = x^2 - s_i$, we may assume that roots of $h(x) - s_i \equiv 0 \bmod p^m$ are of the form

$$-a + t,\ -a - t = -(-a+t) - 2a \bmod p^m.$$

Therefore, we may assume that the roots are

$$r_{i,1} \text{ and } r_{i,2} := p^m - r_{i,1} - 2a \quad (0 \le r_{i,1} \le p^m - 1).$$

Let us show that

$$0 \le r_{i,2} = p^m - r_{i,1} - 2a \le p^m - 1 \tag{4.1}$$

if p^m is sufficiently large. If $p^m - r_{i,1} - 2a \ge p^m$, hence $0 \le r_{i,1} \le -2a$ holds for infinitely many p^m, then there is an integer R with $0 \le R \le -2a$ such that $R = r_{i,1}$ for infinitely many numbers p^m. This R satisfies $f(R) \equiv \prod_i (h(r_{i,1}) - s_i) \equiv 0 \bmod p^m$ for infinitely many p^m, which implies $f(R) = 0$. Thus we have a contradiction, and hence $p^m - r_{i,1} - 2a \le p^m - 1$ for sufficiently large p^m.
If $p^m - r_{i,1} - 2a < 0$ for infinitely many p^m, then we have $-2a < r_{i,1} - p^m \le -1$, and hence there is an integer R' such that $-2a < R' \le -1$ and $R' = r_{i,1} - p^m$ for infinitely many p^m. This R' satisfies $f(R') \equiv f(r_{i,1}) \equiv 0 \bmod p^m$ for infinitely many p^m, which implies the contradiction $f(R') = 0$. Thus we have shown (4.1) and then roots $r_{i1}, r_{i,2}$ of $h(x) - s_i \equiv 0 \bmod p^m$

are in the interval $[0, p^m - 1]$ for sufficiently large p^m. Hence we have

$$\begin{aligned} k_{p^m} p^m &= a_{n-1} + \sum_{i=1}^{n/2} \sum_{j=1}^{2} r_{i,j} \\ &= an + n/2 \cdot (p^m - 2a) \\ &= n/2 \cdot p^m. \end{aligned}$$

Thus we have $k_{p^m} = n/2$.

Next, we consider the case that $h(x) = (x+a)(x+a+1) = x^2 + (2a+1)x + a(a+1)$ for an integer a, hence $f(x) = x^n + n/2 \cdot (2a+1)x^{n-1} + \ldots$. Since a transformation $x \to -x-1-2a$ of order 2 acts on the set of roots of $h(x) - s_i \equiv 0 \bmod p^m$, we may assume that $r_{i,1}, r_{i,2} := p^m - r_{i,1} - 1 - 2a$ are the roots of $h(x) - s_i \equiv 0 \bmod p^m$. We take $r_{i,1}$ so that $0 \le r_{i,1} \le p^m - 1$. Let us see that $0 \le r_{i,2} \le p^m - 1$, i.e. $0 \le p^m - r_{i,1} - 1 - 2a \le p^m - 1$ if p^m is sufficiently large.
If $p^m - r_{i,1} - 1 - 2a < 0$ for infinitely many p^m, then we have $-2a \le p^m - r_{i,1} - 1 - 2a < 0$. Therefore there is an integer R such that $-2a \le R < 0$ and $R = p^m - r_{i,1} - 1 - 2a$ for infinitely many p^m, which means $h(r_{i,2}) - s_i \equiv h(R) - s_i \equiv 0 \bmod p^m$, hence $f(R) \equiv 0 \bmod p^m$. This contradicts that $f(x)$ is irreducible. Thus we have $0 \le p^m - r_{i,1} - 1 - 2a \,(= r_{i,2})$ if p^m is sufficiently large. Suppose that $p^m - r_{i,1} - 1 - 2a \ge p^m$ for infinitely many p^m, which implies $0 \le r_{i,1} \le -1-2a$; then there is an integer R' with $0 \le R' \le -1-2a$ such that $r_{i,1} = R'$, hence $h(R') - s_i \equiv 0 \bmod p^m$ for infinitely many p^m. This is also a contradiction. Thus we conclude that $r_{i,1}, p^m - r_{i,1} - 1 - 2a$ are all roots of $h(x) - s_i \equiv 0 \bmod p^m$ in the interval $[0, p^m - 1]$ if p^m is large. Hence we have

$$\begin{aligned} k_{p^m} p^m &= a_{n-1} + \sum_{i=1}^{n/2} \sum_{j=1}^{2} r_{i,j} \\ &= n/2 \cdot (2a+1) + \sum_{i=1}^{n/2} (r_{i,1} + (p^m - r_{i,1} - 1 - 2a)) \\ &= n/2 \cdot (2a+1) + n/2 \cdot (p^m - 1 - 2a) \\ &= n/2 \cdot p^m. \end{aligned}$$

Thus we have $k_{p^m} = n/2$. □

Proof of Proposition 2.1. The equation (0.4) implies $R_f = a_{n-1} + \sum R_i \equiv a_{n-1} + \sum r_{i,p^m} = k_{p^m} p^m \bmod L$, i.e.

$$R_f \equiv k_{p^m} p^m \equiv k_{p^m} q^m \bmod L$$

with $1 \leq {}^\exists k_{p^m} \leq n-1$ by Proposition 0.1. So, we have $d = (R_f, L) = (k_{p^m}, L)$ by $(q, L) = 1$. We have only to put $K = k_{p^m}$. Since p decomposes completely on $\mathbb{Q}(f)$, hence on $\mathbb{Q}(f) \cap \mathbb{Q}(\zeta_{L/d})$, and $p \equiv q \bmod L$, we have $[[q]] = [[p]] = [[1]]$ on $\mathbb{Q}(f) \cap \mathbb{Q}(\zeta_{L/d})$. □

Proof of Proposition 2.2. We note that $Spl(f)$ is the set of all prime numbers. Suppose that $a > 0$ first; then a local solution r with $0 \leq r < p^m$ of $f(x) = x + a \equiv 0 \bmod p^m$ is given by $p^m - a$ if $p^m > a$, and the condition $r \equiv R \bmod L$ is equivalent to $p^m \equiv R + a \bmod L$. Hence, if the density is non-zero, then $R + a \equiv l_0^m \bmod L$ for an integer l_0 with $(l_0, L) = 1$. Since the set of primes p satisfying $p^m \equiv R + a \equiv l_0^m \bmod L$ is the set $\{p \mid p \equiv l \bmod L$ for some integer l satisfying $l^m \equiv l_0^m \bmod L\}$, the density of primes p with $p^m \equiv R + a \bmod L$ is equal to $\#\{q \bmod L \mid q^m \equiv 1 \bmod L\}/\varphi(L)$ by Dirichlet's prime density theorem on an arithmetic progression. If $a \leq 0$, then the local solution is $-a$ for a large p^m and the condition $r \equiv R \bmod L$ is equivalent to $R + a \equiv 0 \bmod L$, hence the assertion is obvious. □

Proof of Proposition 2.3. Since $pr(f, m, L, \{R_i\})$ depends not on $\{R_i\}$ itself but on R_f, and R_f is determined through $R_f \equiv Kq^m \bmod L$ by K, q, the sum of $pr(f, m, L, \{R_i\})$ on $\{R_i\}$ is equal to

$$\sum_{K,q} E_n(K) \cdot \frac{1}{[\mathbb{Q}(\zeta_L) : \mathbb{Q}(f) \cap \mathbb{Q}(\zeta_{L/d})]}$$

where we put $d = (K, L) \leq n-1$, and K, q satisfy the conditions (2.5.1), (2.5.3)

$$= \sum_{\substack{d=1,\\ d \mid L}}^{n-1} \sum_{\substack{K=1,\\ (K,L)=d}}^{n-1} \sum_{q} E_n(K) \cdot \frac{1}{[\mathbb{Q}(\zeta_L) : \mathbb{Q}(f) \cap \mathbb{Q}(\zeta_{L/d})]}$$

where q satisfies conditions $q \bmod L$, that is $(q, L) = 1, [[q]] = [[1]]$ on $\mathbb{Q}(f) \cap \mathbb{Q}(\zeta_{L/d})$,

$$
\begin{aligned}
&= \sum_{\substack{d=1,\\ d \mid L}}^{n-1} \sum_{\substack{K=1,\\ (K,L)=d}}^{n-1} E_n(K) \cdot \frac{1}{[\mathbb{Q}(\zeta_L) : \mathbb{Q}(f) \cap \mathbb{Q}(\zeta_{L/d})]} \times \\
&\qquad\qquad \# \left\{ q \bmod L \,\middle|\, \begin{array}{l} (q, L) = 1, \\ [[q]] = [[1]] \text{ on } \mathbb{Q}(f) \cap \mathbb{Q}(\zeta_{L/d}) \end{array} \right\} \\
&= \sum_{\substack{d=1,\\ d \mid L}}^{n-1} \sum_{\substack{K=1,\\ (K,L)=d}}^{n-1} E_n(K) \\
&= \sum_{K=1}^{n-1} E_n(K) \\
&= 1.
\end{aligned}
$$

□

Corrections to [8]:

(1) In case that f of degree 4 is irreducible and decomposable, we gave a conjecture "If T is odd, ..." from the sixth line from the bottom to the second line in the section 2.2.1 on p.28 in [8]. It is false. There are other types of densities. Data were insufficient. Details will be given in a subsequent paper.

(2) As stated just after Expectation 2, the main conjecture on p.25 is false for polynomials of degree 6 given after Conjecture 1.

References

1. W. Duke, J.B. Friedlander and H. Iwaniec, Equidistribution of roots of a quadratic congruence to prime moduli, Ann. of Math. (2) **141** (1995), 423-441.
2. W. Feller, *An Introduction to Probability Theory and Its Applications*, Vol. 2, J. Wiley, New York, 1966.
3. T. Hadano, Y. Kitaoka, T. Kubota, M. Nozaki, Densities of sets of primes related to decimal expansion of rational numbers, Number Theory: Tradition and Modernization, pp. 67-80, W. Zhang and Y. Tanigawa, eds. 2006 Springer Science + Business Media,Inc.
4. Y. Kitaoka, A statistical relation of roots of a polynomial in different local fields, Math. Comp. **78** (2009), 523-536.
5. Y. Kitaoka, A statistical relation of roots of a polynomial in different local fields II, Number Theory: Dreaming in Dreams (Series on Number Theory and Its Application Vol. 6), pp. 106-126, World Scientific, 2010.

6. Y. Kitaoka, A statistical relation of roots of a polynomial in different local fields III, Osaka J. Math. **49** (2012), 393-420.
7. Y. Kitaoka, *Introduction to Algebra* (in Japanese), Kin-en-Shobo (2012).
8. Y. Kitaoka, A statistical relation of roots of a polynomial in different local fields IV, Uniform Distribution Theory **8** (2013), no.1, 17-30.
9. A. Schinzel, Polynomials with Special Regard to Reducibility, Cambridge University Press (2000).
10. Á. Tóth, Roots of quadratic congruences, Internat. Math. Res. Notices (2000), 719-739.

A SURVEY ON THE THEORY OF UNIVERSALITY FOR ZETA AND L-FUNCTIONS

KOHJI MATSUMOTO

Graduate School of Mathematics, Nagoya University,
Chikusa-ku, Nagoya 464-8602, Japan
E-mail: kohjimat@math.nagoya-u.ac.jp

1. Voronin's universality theorem

Let $\mathbb{N}$ be the set of positive integers, $\mathbb{N}_0 = \mathbb{N} \cup \{0\}$, $\mathbb{Z}$ the ring of rational integers, $\mathbb{Q}$ the field of rational numbers, $\mathbb{R}$ the field of real numbers, and $\mathbb{C}$ the field of complex numbers. In the present article, the letter p denotes a prime number.

For any open region $D \subset \mathbb{C}$, denote by $H(D)$ the space of $\mathbb{C}$-valued holomorphic functions defined on D, equipped with the topology of uniform convergence on compact sets.

For any subset $K \subset \mathbb{C}$, let $H^c(K)$ be the set of continuous functions defined on K, and are holomorphic in the interior of K, and let $H_0^c(K)$ be the set of all elements of $H^c(K)$ which are non-vanishing on K. By meas S we mean the usual Lebesgue measure of the set S, and by $\#S$ the cardinality of S.

Let $s = \sigma + it \in \mathbb{C}$ (where $\sigma = \Re s$, $t = \Im s$, and $i = \sqrt{-1}$), and $\zeta(s)$ the Riemann zeta-function. This function is defined by the infinite series $\zeta(s) = \sum_{n=1}^{\infty} n^{-s}$ in the half-plane $\sigma > 1$, and can be continued meromorphically to the whole of $\mathbb{C}$. It is well known that the investigation of $\zeta(s)$ in the critical strip $0 < \sigma < 1$ is extremely important in number theory, but its behaviour there still remains quite mysterious. A typical example of expressing one of such mysterious features of $\zeta(s)$ is Voronin's *universality theorem.*

Consider the closed disc $K(r)$ with center $3/4$ and radius r, where $0 < r < 1/4$. Then Voronin [205] proved that for any $f \in H_0^c(K(r))$ and any

$\varepsilon > 0$, there exists a positive number τ for which

$$\max_{s \in K(r)} |\zeta(s + i\tau) - f(s)| < \varepsilon \tag{1.1}$$

holds. Roughly speaking, *any non-vanishing holomorphic function can be approximated uniformly by a certain shift of* $\zeta(s)$.

Actually, Voronin's proof essentially includes the fact that the set of such τ has a positive lower density. Moreover, now it is known that the disc can be replaced by more general type of sets. The modern statement of Voronin's theorem is as follows. Let $D(a,b)$ $(a < b)$ indicate a vertical strip

$$D(a,b) = \{s \in \mathbb{C} \mid a < \sigma < b\}.$$

Theorem 1.1. *(Voronin's universality theorem) Let K be a compact subset of $D(1/2,1)$ with connected complement, and $f \in H_0^c(K)$. Then, for any $\varepsilon > 0$,*

$$\liminf_{T \to \infty} \frac{1}{T} \text{meas} \left\{ \tau \in [0,T] \;\middle|\; \sup_{s \in K} |\zeta(s + i\tau) - f(s)| < \varepsilon \right\} > 0 \tag{1.2}$$

holds.

Let $\varphi(s)$ be a Dirichlet series, and let K be a compact subset of $D(a,b)$ with connected complement. If

$$\liminf_{T \to \infty} \frac{1}{T} \text{meas} \left\{ \tau \in [0,T] \;\middle|\; \sup_{s \in K} |\varphi(s + i\tau) - f(s)| < \varepsilon \right\} > 0 \tag{1.3}$$

holds for any $f \in H_0^c(K)$ and any $\varepsilon > 0$, then we say that the *universality* holds for $\varphi(s)$ in the region $D(a,b)$. Theorem 1.1 implies that the universality holds for the Riemann zeta-function in the region $D(1/2,1)$.

The Riemann zeta-function has the Euler product expression $\zeta(s) = \prod_p (1 - p^{-s})^{-1}$, where p runs over all prime numbers. This is valid only in the region $\sigma > 1$, but even in the region $D(1/2,1)$, it is possible to show that a finite truncation of the Euler product "approximates" $\zeta(s)$ in a certain mean-value sense. On the other hand, since $\{\log p\}_p$ is linearly independent over $\mathbb{Q}$, we can apply the Kronecker-Weyl approximation theorem to obtain that any target function $f(s)$ can be approximated by the above finite truncation. This is the basic structure of the proof of Theorem 1.1.

Remark 1.1. Here we recall the statement of the Kronecker-Weyl theorem. For $x \in \mathbb{R}$, the symbol $||x||$ stands for the distance from x to the nearest integer. Let $\alpha_1, \ldots, \alpha_m$ be real numbers, linearly independent over $\mathbb{Q}$. Then, for any real numbers $\theta_1, \ldots, \theta_m$ and any $\varepsilon > 0$,

$$\lim_{T \to \infty} \frac{1}{T} \text{meas}\{\tau \in [0, T] \mid ||\tau\alpha_k - \theta_k|| < \varepsilon \ (1 \leq k \leq m)\} > 0 \tag{1.4}$$

holds.

So far, three proofs are known for Theorem 1.1. Needless to say, one of them is Voronin's original proof, which is also reproduced in [64]. This proof is based on, besides the above facts, Pecherskiĭ's rearrangement theorem [185] in Hilbert spaces. The second proof is given by Good [41], which will be discussed in Section 16. Gonek was inspired by the idea of Good to write his thesis [40], in which he gave a modified version of Good's argument. The third is a more probabilistic proof due to Bagchi [3] [4]. Bagchi [3] is an unpublished thesis, but its contents are carefully expounded in [76]. A common feature of the work of Gonek and Bagchi is that they both used the approximation theorem of Mergelyan [141] [142].

Remark 1.2. Mergelyan's theorem asserts that, when K is a compact subset of $\mathbb{C}$ with connected complement, any $f \in H^c(K)$ can be approximated by polynomials uniformly on K. This is a complex analogue of the classical approximation theorem of Weierstrass.

Here we mention some pre-history. In 1914, Bohr and Courant [13] proved that, for any σ satisfying $1/2 < \sigma \leq 1$, the set

$$\{\zeta(\sigma + i\tau) \mid \tau \in \mathbb{R}\}$$

is dense in $\mathbb{C}$. In the next year, Bohr [11] proved that the same result holds for $\log \zeta(\sigma + i\tau)$. These results are called *denseness theorems.*

Before obtaining his universality theorem, Voronin [204] discovered the following multi-dimensional analogue of the theorem of Bohr and Courant.

Theorem 1.2. *(Voronin [204]) For any σ satisfying $1/2 < \sigma \leq 1$, the set*

$$\{(\zeta(\sigma + i\tau), \zeta'(\sigma + i\tau), \ldots, \zeta^{(m-1)}(\sigma + i\tau)) \mid \tau \in \mathbb{R}\}$$

(where $\zeta^{(k)}$ denotes the k-th derivative) is dense in $\mathbb{C}^m$.

Remark 1.3. Actually Voronin proved a stronger result. For any $s = \sigma + it$ with $1/2 < \sigma \leq 1$ and for any $h > 0$, the set

$$\{(\zeta(s + inh), \zeta'(s + inh), \ldots, \zeta^{(m-1)}(s + inh)) \mid n \in \mathbb{N}\}$$

is dense in $\mathbb{C}^m$.

The universality theorem of Voronin may be regarded as a natural next step, because it is a kind of infinite-dimensional analogue of the theorem of Bohr and Courant, or a *denseness theorem in a function space.*

Another refinement of the denseness theorem of Bohr and Courant is the *limit theorem* on $\mathbb{C}$, due to Bohr and Jessen [14]. Recently, this theorem is usually formulated by probabilistic terminology. Let $\sigma > 1/2$. For any Borel subset $A \subset \mathbb{C}$, put

$$P_{T,\sigma}(A) = \frac{1}{T}\text{meas}\{\tau \in [0,T] \mid \zeta(\sigma + i\tau) \in A\}.$$

This $P_{T,\sigma}$ is a probability measure on $\mathbb{C}$. Then the modern formulation of the limit theorem of Bohr and Jessen is that there exists a probability measure P_σ, to which $P_{T,\sigma}$ is convergent weakly as $T \to \infty$ (see [76, Chapter 4]).

Bagchi [3] proved an analogue of the above theorem of Bohr and Jessen on a certain function space, and used it in his alternative proof of the universality theorem. Therefore, to prove some universality-type theorem by Bagchi's method, it is necessary to obtain some *functional limit theorem* similar to that of Bagchi. There are indeed a lot of papers devoted to the proofs of various functional limit theorems, for the purpose of showing various universality theorems. However in the following sections we do not explicitly mention this closely related topic of functional limit theorems, and for the details we refer to Laurinčikas [76] or J. Steuding [201]. The connection between the theory of universality and functional limit theorems is also discussed in the author's survey articles [136] [137].

After the publication of Voronin's theorem in 1975, now almost forty years have passed. Voronin's theorem attracted a lot of mathematicians, and hence, after Voronin, quite many papers on universality theory have been published. The aim of the present article is to survey the developments in this theory in these forty years. The developments in these years can be divided into three stages.

(I) The first stage: 1975 ~ 1987.

(II) The second stage: 1996 ~ 2007.

(III) The third stage: 2007 ~ present.

In the next section we will give a brief discussion what were the main topics in each of these stages.

The author expresses his sincere gratitude to Professors R. Garunkštis, R. Kačinskaitė, S. Kanemitsu, E. Karikovas, A. Laurinčikas, R. Macaitienė, H. Mishou, H. Nagoshi, T. Nakamura, Ł. Pańkowski, D. Šiaučiūnas, and J.

Steuding for their valuable comments and/or sending relevant articles.

2. A rough sketch of the history

(I) The first stage.

This is the first decade after Voronin's paper. The original impact of Voronin's discovery was still fresh. Mathematicians who were inspired by Voronin's paper tried to discuss various generalizations, analogies, refinements and so on. Here is the list of main results obtained in this decade.

- Alternative proofs (mentioned in Section 1).
- Generalizations to the case of Dirichlet L-functions, Dedekind zeta-functions etc.
- The joint universality.
- The strong universality (for Hurwitz zeta-functions).
- The strong recurrence.
- The discrete universality.
- The χ-universality.
- The hybrid universality.
- A quantitative result.

It is really amazing that many important aspects in universality theory, developed extensively in later decades, had already been introduced and studied in this first decade. Unfortunately, however, many of those results were written only in the theses of Voronin [207], Gonek [40] and Bagchi [3], all of them remain unpublished. This situation is probably one of the reasons why in the next several years there were so few publications on universality.

(II) The second stage.

During several years around 1990, the number of publications concerning universality is very small. Of course it is not completely empty. For example, the book of Karatsuba and Voronin [64] was published in this period. But the author prefers to choose the year 1996 as the starting point of the second stage, because in this year the important book [76] of Laurinčikas was published. This is the first textbook which is mainly devoted to the theory of universality and related topics, and especially, provides the details of unpublished work of Bagchi. Thanks to the existence of this book, many mathematicians of younger generation can now easily go into the theory of universality. In fact, in this second stage, a lot of students of Laurinčikas started to publish their papers, and they formed the strong Lithuanian school.

The main topic in this decade was probably the attempt to extend the class of zeta and L-functions for which the universality property holds. It is now known that the universality property is valid for a rather wide class of zeta-functions. J. Steuding's lecture note [201] includes the exposition of this result, and also of many other results obtained after the publication of the book of Laurinčikas [76]. Therefore the publication year of this lecture note is appropriate to the end of the second stage.

(III) The third stage.

Now comes the third, present stage. The theory of universality is now developing into several new directions. The notions of

- the mixed universality,
- the composite universality,
- the ergodic universality,

were introduced recently. Other topics in universality theory have also been discussed extensively.

In the following sections, we will discuss more closely each topic in universality theory.

3. Generalization to zeta and L-functions with Euler products

Is it possible to prove the universality property for other zeta and L-functions? This is surely one of the most fundamental question. In Section 1 we explained that a key point in the proof of Theorem 1.1 is the Euler product expression. Therefore we can expect the universality property for other zeta and L-functions which have Euler products.

The universality of the following zeta and L-functions were proved in the first decade.

- Dirichlet L-functions $L(s,\chi) = \sum_{n=1}^{\infty} \chi(n)n^{-s}$ (χ is a Dirichlet character; see Voronin [205]) in $D(1/2,1)$,
- certain Dirichlet series with multiplicative coefficients (see Reich [188], Laurinčikas [68] [69] [70] [71] [72]),
- Dedekind zeta-functions $\zeta_F(s) = \sum_{\mathfrak{a}} N(\mathfrak{a})^{-s}$, where F is a number field, $\mathfrak{a}$ denotes a non-zero integral ideal, and $N(\mathfrak{a})$ its norm (see Voronin [207] [208], Gonek [40], Reich [189] [190]). Here, the universality can be proved in $D(1/2,1)$ if F is an Abelian extension of $\mathbb{Q}$ (Gonek [40]), but for general $F(\neq \mathbb{Q})$, the proof is valid only in the narrower region $D(1-d_F^{-1},1)$, where $d_F = [F:\mathbb{Q}]$. The reason is that the mean value estimate for $\zeta_F(s)$, applicable to the proof of universality, is known at present only in $D(1-d_F^{-1},1)$.

Later, Laurinčikas also obtained universality theorems for Matsumoto zeta-functions* ([78] under a strong assumption), and for the zeta-function attached to Abelian groups ([82] [83]). Laurinčikas and Šiaučiūnas [122] proved the universality for the periodic zeta-function $\zeta(s, \mathfrak{A}) = \sum_{n=1}^{\infty} a_n n^{-s}$ (where $\mathfrak{A} = \{a_n\}_{n=1}^{\infty}$ is a multiplicative periodic sequence of complex numbers) when the technical condition

$$\sum_{m=1}^{\infty} |a_{p^m}| p^{-m/2} \leq c \tag{3.1}$$

(with a certain constant $c < 1$) holds for any prime p. Schwarz, R. Steuding and J. Steuding [192] proved another universality theorem on certain general Euler products with conditions on the asymptotic behaviour of coefficients.

However, there was an obstacle when we try to generalize further. A typical class of L-functions with Euler products is that of automorphic L-functions. Let $g(z) = \sum_{n=1}^{\infty} a(n) e^{2\pi i n z}$ be a holomorphic normalized Hecke-eigen cusp form of weight κ with respect to $SL(2, \mathbb{Z})$ and let $L(s, g) = \sum_{n=1}^{\infty} a(n) n^{-s}$ be the associated L-function. The universality of $L(s, g)$ was first discussed by Kačėnas and Laurinčikas [53], but they showed the universality only under a very strong assumption.

What was the obstacle? The asymptotic formula

$$\sum_{p \leq x} \frac{1}{p} = \log\log x + a_1 + O(\exp(-a_2 \sqrt{\log x})) \tag{3.2}$$

(where a_1, a_2 are constants) is classically well known, and is used in the proof of Theorem 1.1. However the corresponding asymptotic formula for the sum $\sum_{p \leq x} |a(p)| p^{-1}$ is not known. To avoid this obstacle, Laurinčikas and the author [111] invented a new method of using (3.2), combined with the known asymptotic formula

$$\sum_{p \leq x} |\widetilde{a}(p)|^2 = \frac{x}{\log x}(1 + o(1)), \tag{3.3}$$

where $\widetilde{a}(p) = a(p) p^{-(\kappa-1)/2}$. This method is called the *positive density method.* Modifying Bagchi's argument by virtue of this positive density method, one can show the following unconditional result.

Theorem 3.1. *(Laurinčikas and Matsumoto [111]) The universality holds for $L(s, g)$ in the region $D(\kappa/2, (\kappa + 1)/2)$.*

---

*This notion was first introduced by the author [133], to which the limit theorem of Bohr and Jessen was generalized. See also Kačinskaitė's survey article [56].

The positive density method was then applied to prove the universality for more general class of L-functions; certain Dirichlet series with multiplicative coefficients (Laurinčikas and Šleževičienė [126]), L-functions attached to new forms with respect to congruence subgroups (Laurinčikas, Matsumoto and J. Steuding [116]), L-functions attached to a cusp form with character (Laurinčikas and Macaitienė [105]), and a certain subclass of the Selberg class† (J. Steuding [197]). J. Steuding extended his result further in his lecture note [201]. He introduced a wide class $\widetilde{S}$ of L-functions defined axiomatically and proved the universality for elements of $\widetilde{S}$. The class $\widetilde{S}$, now sometimes called the Steuding class, is not included in the Selberg class, but is a subclass of the class of Matsumoto zeta-functions.

Since the Shimura-Taniyama conjecture has been established, we now know that the L-function $L(s, E)$ attached to a non-singular elliptic curve E over $\mathbb{Q}$ is an L-function attached to a new form. Therefore the universality for $L(s, E)$ is included in [116]. The universality of positive powers of $L(s, E)$ was studied in Garbaliauskienė and Laurinčikas [31].

Mishou [144] [145] used a variant of the positive density method to show the universality for Hecke L-functions of algebraic number fields in the region $D(1 - d_F^{-1}, 1)$. Lee [127] showed that, under the assumption of a certain density estimate of the number of zeros, it is possible to prove the universality for Hecke L-functions in the region $D(1/2, 1)$. The universality for Artin L-functions was proved by Bauer [7] by a different method, based on Voronin's original idea.

Let g_1 and g_2 be cusp forms. The universality for the Rankin-Selberg L-function $L(s, g_1 \otimes g_1)$ was shown by the author [134], and for $L(s, g_1 \otimes g_2)$ $(g_1 \neq g_2)$ was by Nagoshi [165] (both in the narrower region $D(3/4, 1)$). The latter proof is based on the above general result of J. Steuding [197] [201]. The universality of symmetric m-th power L-functions $(m \leq 4)$ and their Rankin-Selberg L-functions was studied by Li and Wu [129].

Another general result obtained by the positive density method is the following theorem, which is an extension of the result of J. Steuding [197].

Theorem 3.2. *(Nagoshi and J. Steuding [166]) Let* $\varphi(s) = \sum_{n=1}^{\infty} a(n)n^{-s}$ *be a Dirichlet series belonging to the Selberg class. Denote the degree of*

†The notion of Selberg class was introduced by Selberg [193]. For basic definitions and results in this theory, consult a survey [63] of Kaczorowski and Perelli, or J. Steuding [201].

$\varphi(s)$ by d_φ, and put $\sigma_\varphi = \max\{1/2, 1-d_\varphi^{-1}\}$. Assume that

$$\sum_{p\leq x} |a(p)|^2 = \frac{x}{\log x}(\lambda + o(1)) \tag{3.4}$$

holds with a certain positive constant λ. Then the universality holds for $\varphi(s)$ in the region $D(\sigma_\varphi, 1)$.

Remark 3.1. When φ is the Riemann zeta-function, the formula (3.4) (with $\lambda = 1$) is nothing but the prime number theorem. Therefore we may say that the positive density method enables us to prove the universality for zeta-functions with Euler products, provided an asymptotic formula of the prime-number-theorem type is known.

Let h be a Hecke-eigen Maass form, and let $L(s,h)$ be the associated L-function. The universality for $L(s,h)$ was proved by Nagoshi [163] [164]. In the proof of Theorem 3.1 in [111], Deligne's estimate (Ramanujan's conjecture) $|\widetilde{a}(p)| \leq 2$ is essentially used. Assumptions of the same type are required in Steuding's general result [197] [201] and also in Theorem 3.2. Since Ramanujan's conjecture for Maass forms has not yet been proved, Nagoshi in his first paper [163] assumed the validity of Ramanujan's conjecture for h to show the universality. Then in the second paper [164] he succeeded to remove this assumption, by invoking the asymptotic formula for the fourth power mean of the coefficients due to M. Ram Murty.

Theorem 3.3. *(Nagoshi [164]) The universality holds for $L(s,h)$ in the region $D(1/2, 1)$.*

Another important class of zeta-functions which have Euler products is the class of Selberg zeta-functions. In this case, instead of the prime-number-theorem type of results, the prime geodesic theorem plays an important role. Let

$$\mathcal{D} = \{d \in \mathbb{N} \mid d \equiv 0 \text{ or } 1 \pmod 4, d \text{ is not a square}\}.$$

For each $d \in \mathcal{D}$, let $h^+(d)$ be the number of inequivalent primitive quadratic forms of discriminant d, and $\varepsilon(d) = (u(d) + v(d)\sqrt{d})/2$, where $(u(d), v(d))$ is the fundamental solution of the Pell equation $u^2 - v^2 d = 4$. Then, the prime geodesic theorem for $SL(2,\mathbb{Z})$ implies

$$\sum_{\substack{d\in\mathcal{D} \\ \varepsilon(d)^2\leq x}} h^+(d) = \int_0^x \frac{dt}{\log t} + O(x^\alpha) \tag{3.5}$$

with a certain $\alpha < 1$.

Theorem 3.4. *(Drungilas, Garunkštis and Kačėnas [23]) Let $Z(s)$ be the Selberg zeta-function attached to $SL(2,\mathbb{Z})$. If (3.5) holds, then the universality holds for $Z(s)$ in the region $D(1/2+\alpha/2,1)$.*

As for the value of α, it is known that one can take $\alpha = 71/102+\varepsilon$ for any $\varepsilon > 0$ (Cai [19]). It is conjectured that one could take $\alpha = 1/2+\varepsilon$. If the conjecture is true, then Theorem 3.4 implies that $Z(s)$ has the universality property in $D(3/4,1)$. The paper [23] includes a discussion which suggests that $D(3/4,1)$ is the widest possible region where the universality for $Z(s)$ holds.

4. The joint universality for zeta and L-functions with Euler products

The results presented in the previous sections give approximation properties of some single zeta or L-function. Here we discuss simultaneous approximations by several zeta or L-functions.

Let $K_1,\ldots,K_r$ be compact subsets of $D(a,b)$ with connected complements, and $f_j \in H_0^c(K_j)$ $(1 \le j \le r)$. If Dirichlet series $\varphi_1(s),\ldots,\varphi_r(s)$ satisfy

$$\liminf_{T\to\infty} \frac{1}{T}\text{meas}\left\{\tau \in [0,T] \;\middle|\; \sup_{s\in K_j} |\varphi_j(s+i\tau) - f_j(s)| < \varepsilon \quad (1 \le j \le r)\right\} > 0 \tag{4.1}$$

for any $\varepsilon > 0$, we call the *joint universality* holds for $\varphi_1(s),\ldots,\varphi_r(s)$ in the region $D(a,b)$. The joint universality for Dirichlet L-functions was already obtained in the first decade by Voronin [206] [207] [208], Gonek [40], and Bagchi [3] [4], independently of each other:

Theorem 4.1. *(Voronin, Gonek, Bagchi) Let $\chi_1,\ldots,\chi_r$ be pairwise non-equivalent Dirichlet characters, and $L(s,\chi_1),\ldots,L(s,\chi_r)$ the corresponding Dirichlet L-functions. Then the joint universality holds for $L(s,\chi_1),\ldots,L(s,\chi_r)$ in the region $D(1/2,1)$.*

To prove such a theorem of simultaneous approximations, it is obviously necessary that the behaviour of L-functions appearing in the theorem should be "independent" of each other. In the situation of Theorem 4.1, this is embodied by the orthogonality relation of Dirichlet characters, which is essentially used in the proof.

The joint universality for Dedekind zeta-functions was studied by Voronin [207] [208]. Bauer's work [7] mentioned in the preceding section actually proves a joint universality theorem on Artin L-functions, in the region $D(1-(2d_F)^{-1}, 1)$. Lee [128] extended the region to $D(1/2, 1)$ under the assumption of a certain density estimate.

Let g_j $(1 \leq j \leq r)$ be multiplicative arithmetic functions. Laurinčikas [74] considered the joint universality of the associated Dirichlet series $\sum_{n=1}^{\infty} g_j(n)n^{-s}$ $(1 \leq j \leq r)$. In this case, the "independence" condition is given by the following matrix condition. Let $P_1, \ldots, P_k$ $(k \geq r)$ be certain sets of prime numbers, with the condition that $\sum_{p \leq x, p \in P_l} p^{-1}$ satisfies a good asymptotic formula, and assume that $g_j(n)$ is a constant g_{jl} on the set P_l $(1 \leq j \leq r, 1 \leq l \leq k)$. Laurinčikas [74] proved a joint universality theorem under the condition that the rank of the matrix $(g_{jl})_{1 \leq j \leq r, 1 \leq l \leq k}$ is equal to r.

Laurinčikas frequently used various matrix conditions to obtain joint universality theorems. A joint universality theorem on Matsumoto zeta-functions under a certain matrix condition was proved in [79]. A joint universality for automorphic L-functions under a certain matrix condition was discussed in [112].

A matrix condition naturally appears in the joint universality theory of periodic zeta-functions (see [107] [103]). Let $\mathfrak{A}_j = \{a_{jn}\}_{n=1}^{\infty}$ be a multiplicative periodic sequence (whose least period we denote by k_j) of complex numbers, and $\zeta(s, \mathfrak{A}_j)$ the associated periodic zeta-function $(1 \leq j \leq r)$. Let k be the least common multiple of $k_1, \ldots, k_r$. Define the matrix $A = (a_{jl})_{j,l}$, where $1 \leq j \leq r$ and $1 \leq l \leq k$, $(l, k) = 1$. Then Laurinčikas and Macaitienė [103] proved the joint universality for $\zeta(s, \mathfrak{A}_1), \ldots, \zeta(s, \mathfrak{A}_r)$ in the region $D(1/2, 1)$, if we assume $\mathrm{rank}(A) = r$ and a technical condition similar to (3.1).

Using the positive density method, it is possible to prove a joint universality theorem for twisted automorphic L-functions. Let $g(z) = \sum_{n=1}^{\infty} a(n)e^{2\pi i n z}$ be a holomorphic normalized Hecke-eigen cusp form, $\chi_1, \ldots, \chi_r$ be pairwise non-equivalent Dirichlet characters, and $L(s, g, \chi_j) = \sum_{n=1}^{\infty} a(n)\chi_j(n)n^{-s}$ the associated χ_j-twisted L-function.

Theorem 4.2. *(Laurinčikas and Matsumoto [113]) The joint universality holds for $L(s, g, \chi_1), \ldots, L(s, g, \chi_r)$ in the region $D(\kappa/2, (\kappa+1)/2)$.*

To prove this result, we need a prime number theorem for $a(p)$ in arithmetic progressions, that is

$$\sum_{\substack{p\leq x \\ p\equiv h (\mathrm{mod}\ q)}} |\widetilde{a}(p)|^2 = \frac{1}{\varphi(q)}\frac{x}{\log x}(1+o(1)), \tag{4.2}$$

where $(h, q) = 1$ and $\varphi(q)$ is Euler's totient function.

J. Steuding [201, Theorem 12.8] generalized Theorem 4.2 to the Steuding class $\widetilde{S}$. A joint version of [126] was given by Šleževičienė [194]. A joint universality theorem on L-functions of elliptic curves, under a certain matrix condition, was given by Garbaliauskienė, Kačinskaitė and Laurinčikas [29].

Let $\varphi_j(s) = \sum_{n=1}^{\infty} a_j(n)n^{-s}$ $(j = 1, 2)$ be elements of the Selberg class. The following orthogonality conjecture of Selberg [193] is well known: if $\varphi_1(s)$, $\varphi_2(s)$ are primitive, then

$$\sum_{p\leq x}\frac{a_1(p)\overline{a_2(p)}}{p} = \begin{cases} \log\log x + O(1) & \text{if} \quad \varphi_1 = \varphi_2, \\ O(1) & \text{otherwise.} \end{cases} \tag{4.3}$$

Inspired by this conjecture, J. Steuding [201, Section 12.5] proposed:

Conjecture 3. *(J. Steuding) Any finite collection of distinct primitive functions in the Selberg class is jointly universal.* ‡

Towards this conjecture, recent progress has been mainly due to Mishou. In [153], Mishou proved the following. Consider two strips $D_1 = D(1/2, 3/4)$ and $D_2 = D(3/4, 1)$. Let K_j be a compact subset of D_j and $f_j \in H_0^c(K_j)$ $(j = 1, 2)$. Then

$$\liminf_{T\to\infty}\frac{1}{T}\mathrm{meas}\Big\{\tau\in[0,T] \ \Big| \ \sup_{s\in K}|\zeta(s+i\tau)-f_1(s)|<\varepsilon, \tag{4.4}$$
$$|\widetilde{L}(s+i\tau, g) - f_2(s)| < \varepsilon\Big\} > 0$$

holds, where $\widetilde{L}(s, g) = \sum_{n=1}^{\infty} a(n)n^{-(\kappa-1)/2}n^{-s}$ with a cusp form $g(z)$ as before. In Mishou [154], this result was generalized to the case of several L-functions belonging to the Selberg class.

The result (4.4) is weaker than the joint universality, because K_1 and K_2 are in the strips disjoint to each other. In [156], Mishou succeeded in removing this restriction to obtain the following theorem. Let g, g_1, g_2 be holomorphic normalized Hecke-eigen cusp forms.

‡J. Steuding also mentioned the expectation that any two functions φ_1, φ_2 in the Selberg class would be jointly universal if and only if $\sum_{p\leq x}\frac{a_1(p)\overline{a_2(p)}}{p} = O(1)$. However H. Nagoshi, and then H. Mishou, pointed out that there are counter examples to this statement.

Theorem 4.3. *(Mishou [156]) (i) $\zeta(s)$ and $\widetilde{L}(s,g)$ are jointly universal in $D(1/2,1)$.*

(ii) If g_1 and g_2 are distinct, then $\widetilde{L}(s,g_1)$ and $\widetilde{L}(s,g_2)$ are jointly universal in $D(1/2,1)$.

(iii) $\zeta(s)$ and $L(s,\mathrm{sym}^2 g)$ are jointly universal in $D(2/3,1)$.

(iv) If g_1 and g_2 are distinct, then $\zeta(s)$ and $L(s,g_1\otimes g_2)$ are jointly universal in $D(3/4,1)$.

(v) $\widetilde{L}(s,g_1)$ and $L(s,g_1\otimes g_2)$ are jointly universal in $D(3/4,1)$.

Remark 4.1. The universality theorem for $L(s,g\otimes g)$ by the author [134] (mentioned in Section 3) was proved in $D(3/4,1)$, but the above theorem of Mishou especially implies that the universality for $L(s,g\otimes g)$ is valid in the wider region $D(2/3,1)$.

A remarkable feature of Mishou's method is that it does not depend on any periodicity of coefficients. His proof is based on orthogonality relations of Fourier coefficients. Theorem 4.3 is a strong support to Conjecture 3.

5. The strong universality

So far we have talked about universality only for zeta and L-functions with Euler products. However already in the first decade, the universality for zeta-functions without Euler products was also studied. Let $0<\alpha\leq 1$. The Hurwitz zeta-function with the parameter α is defined by $\zeta(s,\alpha)=\sum_{n=0}^{\infty}(n+\alpha)^{-s}$, and does not have the Euler product (except for the special cases $\alpha=1,1/2$). The known universality theorem for $\zeta(s,\alpha)$ is as follows.

Theorem 5.1. *(Bagchi [3], Gonek [40]) Let K be a compact subset of $D(1/2,1)$ with connected complement, and $f\in H^c(K)$. Then for any $\varepsilon>0$,*

$$\liminf_{T\to\infty}\frac{1}{T}\mathrm{meas}\left\{\tau\in[0,T]\;\middle|\;\sup_{s\in K}|\zeta(s+i\tau,\alpha)-f(s)|<\varepsilon\right\}>0 \tag{5.1}$$

holds, provided α is transcendental or rational ($\neq 1,1/2$).

To prove this theorem, when α is transcendental, we use the fact that the elements of the set

$$\{\log(n+\alpha)\mid n\in\mathbb{N}_0\}$$

are linearly independent over $\mathbb{Q}$. On the other hand, when $\alpha=a/b$ is rational, then in view of the formula

$$\zeta(s,a/b)=\frac{b^s}{\varphi(b)}\sum_{\chi(\mathrm{mod}\ b)}\overline{\chi}(a)L(s,\chi), \tag{5.2}$$

we can reduce the problem to the joint universality of Dirichlet L-functions, so we can apply Theorem 4.1. The case of algebraic irrational α is still open.

A remarkable point is that, in the statement of Theorem 5.1, we do not assume that the target function $f(s)$ is non-vanishing on K. This is a big difference from Theorem 1.1, and when a universality-type theorem holds without the non-vanishing assumption, we call it a *strong universality theorem.*

Strong universality has an important application to the theory of zero-distribution. Let $a < \sigma_1 < \sigma_2 < b$, and let

$$N(t; \sigma_1, \sigma_2; \varphi) = \#\{\rho \in \mathbb{C} \mid \sigma_1 \leq \Re\rho \leq \sigma_2, 0 \leq \Im\rho \leq T, \varphi(\rho) = 0\}$$

for a function φ. (In the above definition, zeros are counted with multiplicity.) Then we have the following consequence.

Theorem 5.2. *If $\varphi(s)$ is strongly universal in the region $D(a, b)$, then there exists a positive constant C for which*

$$N(T; \sigma_1, \sigma_2; \varphi) \geq CT \tag{5.3}$$

holds for any σ_1, σ_2 satisfying $a < \sigma_1 < \sigma_2 < b$.

Proof. Let δ be a small positive number, $0 < \varepsilon < \delta$, $\sigma_1 < \sigma_0 < \sigma_2$, and $K = \{s \in \mathbb{C} \mid |s - \sigma_0| < \delta\}$. We choose δ so small that $K \subset D(a, b)$. We apply the strong universality to this K, $f(s) = s - \sigma_0$ and ε to obtain that the set of real numbers τ such that

$$\sup_{|s-\sigma_0|\leq\delta} |\varphi(s + i\tau) - f(s)| < \varepsilon$$

is of positive lower density. Then, for such τ,

$$\sup_{|s-\sigma_0|\leq\delta} |\varphi(s + i\tau) - f(s)| < \delta = \inf_{|s-\sigma_0|=\delta} |f(s)|.$$

Therefore by Rouché's theorem we see that $f(s)+(\varphi(s+i\tau)-f(s)) = \varphi(s+i\tau)$ has the same number of zeros as that of $f(s)$ in the region $|s - \sigma_0| < \delta$, but the latter is obviously 1. That is, for each τ in the above set, $\varphi(s)$ has one zero in $|s - (\sigma_0 + i\tau)| < \delta$. □

Corollary 5.1. *If α is transcendental or rational ($\neq 1, 1/2$), then*

$$C_1 T \leq N(T; \sigma_1, \sigma_2; \zeta(s, \alpha)) \leq C_2 T$$

holds for any $1/2 < \sigma_1 < \sigma_2 < 1$.

As for the upper bound part of this corollary, see [102, Chapter 8, Theorem 4.10].

Further topics on the application of universality to the distribution of zeros will be discussed in Section 9 and Section 15.

Now strong universality theorems are known for many other zeta-functions. The Estermann zeta-function is defined by

$$E\left(s, \frac{k}{l}, \alpha\right) = \sum_{n=1}^{\infty} \sigma_\alpha(n) \exp\left(2\pi i \frac{k}{l} n\right) n^{-s}, \tag{5.4}$$

where k and l are coprime integers and $\sigma_\alpha(n) = \sum_{d|n} d^\alpha$. The strong universality for $E(s, k/l, \alpha)$ was studied in Garunkštis, Laurinčikas, Šleževičienė and J. Steuding [36]. The method is to write $E(s, k/l, \alpha)$ as a linear combination of $E(s, \chi, \alpha) = \sum_{n=1}^{\infty} \sigma_\alpha(n)\chi(n)n^{-s}$, and apply a joint universality theorem for $E(s, \chi, \alpha)$ which follows from Šleževičienė [194].

The Lerch zeta-function is defined by

$$\zeta(s; \alpha, \lambda) = \sum_{n=0}^{\infty} e^{2\pi i \lambda n}(n+\alpha)^{-s},$$

where $0 < \alpha \leq 1$ and λ is real. When λ is an integer, then $\zeta(s; \alpha, \lambda)$ reduces to the Hurwitz zeta-function, so we may assume $0 < \lambda < 1$. The strong universality for $\zeta(s; \alpha, \lambda)$ was proved by Laurinčikas [77] when α is transcendental. The case when α is rational was discussed by Laurinčikas [80]. See also the textbook [102] of Laurinčikas and Garunkštis.

Let $\mathfrak{B} = \{b_n\}$ is a periodic sequence, not necessarily multiplicative. The universality for periodic zeta-functions $\zeta(s, \mathfrak{B}) = \sum_{n=1}^{\infty} b_n n^{-s}$ was first studied by J. Steuding [198] (see [201, Chapter 11]). Kaczorowski [60] proved that there exists a constant $c_0 = c_0(\mathfrak{B})$ such that the universality holds for $\zeta(s, \mathfrak{B})$, provided that

$$\max_{s\in K} \Im(s) - \min_{s \in K} \Im(s) \leq c_0.$$

This result is a consequence of the hybrid joint universality theorem of Kaczorowski and Kulas [61] (see Section 15). Javtokas and Laurinčikas [49] [50] studied the strong universality for periodic Hurwitz zeta-function

$$\zeta(s, \alpha, \mathfrak{B}) = \sum_{n=0}^{\infty} b_n (n+\alpha)^{-s}.$$

They proved that the strong universality holds for $\zeta(s, \alpha, \mathfrak{B})$, when α is transcendental.

A more general situation was considered by Laurinčikas, Schwarz and J. Steuding [121]. Let $\{\lambda_n\}_{n=1}^{\infty}$ be an increasing sequence of real numbers, linearly independent over $\mathbb{Q}$, and $\lambda_n \to \infty$ as $n \to \infty$. Define the general Dirichlet series $f(s) = \sum_{n=1}^{\infty} a_n \exp(-\lambda_n s)$, which is assumed to be convergent absolutely in the region $\sigma > \sigma_a$. Put $r(x) = \sum_{\lambda_n \leq x} 1$ and $c_n = a_n \exp(-\lambda_n \sigma_a)$. We suppose

(i) $f(s)$ cannot be represented as an Euler product,

(ii) $f(s)$ can be continued meromorphically to $\sigma > \sigma_1$, and holomorphic in $D(\sigma_1, \sigma_a)$,

(iii) For $\sigma > \sigma_1$ it holds that $f(s) = O(|t|^{\alpha})$ with some $\alpha > 0$,

(iv) For $\sigma > \sigma_1$ it holds that

$$\int_{-T}^{T} |f(\sigma + it)|^2 dt = O(T),$$

(v) $r(x) = Cx^{\kappa} + O(1)$ with a $\kappa > 1$,

(vi) $|c_n|$ is bounded and $\sum_{\lambda_n \leq x} |c_n|^2 = \theta r(x)(1 + o(1))$ with a $\theta > 0$.

Then we have

Theorem 5.3. *(Laurinčikas, Schwarz and J. Steuding [121]) If $f(s)$ satisfies all the above conditions, then the strong universality holds for $f(s)$ in the region $D(\sigma_1, \sigma_a)$.*

In Section 3 and Section 5, we have seen a lot of examples of zeta and L-functions, with or without Euler products, for which the universality property holds. How general is this property expected to hold? The following conjecture predicts that any "reasonable" Dirichlet series would satisfy the universality property.

Conjecture 4. *(Yu. V. Linnik and I. A. Ibragimov) All functions given by Dirichlet series and meromorphically continuable to the left of the half-plane of absolute convergence are universal in some suitable region.*

Remark 5.1. Actually this conjecture has trivial counter-examples. For example, let $a_n = 1$ if n is a power of 2 and $a_n = 0$ otherwise. The series $\sum_{n=1}^{\infty} a_n n^{-s}$ can be continued to $(2^s - 1)^{-1}$, which is obviously not universal. Therefore some additional condition should be added to make the rigorous statement of the above conjecture.

6. The joint strong universality

The joint universality property is also possible to be valid among zeta-functions without Euler products. The first attempt to this direction is a series of papers of Laurinčikas and the author [110] [114] [115] on the joint universality for Lerch zeta-functions. Here, a matrix condition again appears. Let $\lambda_1, \ldots, \lambda_r$ be rational numbers. Write $\lambda_j = a_j/q_j$, $(a_j, q_j) = 1$, and let k be the least common multiple of $q_1, \ldots, q_r$. Define the matrix $L = (\exp(2l\pi i\lambda_j))_{1\leq l\leq k, 1\leq j\leq r}$. Then, by virtue of a variant of the positive density method, we have

Theorem 6.1. *(Laurinčikas and Matsumoto [110] [114]) Suppose that $\alpha_1, \ldots, \alpha_r$ are algebraically independent over $\mathbb{Q}$, and that* $\mathrm{rank}(L) = r$. *Then the joint strong universality holds for $\zeta(s, \alpha_1, \lambda_1), \ldots, \zeta(s, \alpha_r, \lambda_r)$ in the region $D(1/2, 1)$.*[§]

The Lerch zeta-function $\zeta(s, \alpha, \lambda)$ with rational λ is a special case of periodic Hurwitz zeta-functions. The joint strong universality of periodic Hurwitz zeta-functions was first studied by Laurinčikas [88] [89]. Let $\mathfrak{B}_j = \{a_{nj}\}_{n=0}^{\infty}$ be periodic sequences with period k_j, k be the least common multiple of $k_1, \ldots, k_r$, and define $B = (a_{jl})_{1\leq j\leq r, 1\leq l\leq k}$.

Theorem 6.2. *(Laurinčikas [89]) If α is transcendental and* $\mathrm{rank}(B) = r$, *then the joint strong universality holds for $\zeta(s, \alpha, \mathfrak{B}_1), \ldots, \zeta(s, \alpha, \mathfrak{B}_r)$ in the region $D(1/2, 1)$.*

Next in [51] [124], the joint universality for $\zeta(s, \alpha_j, \mathfrak{B}_j)$ $(1 \leq j \leq r)$ was discussed. Some matrix conditions were still assumed in [51], but finally in [124], a joint universality theorem free from any matrix condition was obtained.

Theorem 6.3. *(Laurinčikas and Skerstonaitė [124]) Assume that the elements of the set*

$$\{\log(n + \alpha_j) \mid 1 \leq j \leq r,\ n \in \mathbb{N}_0\} \tag{6.1}$$

are linearly independent over $\mathbb{Q}$. Then the joint strong universality holds for $\zeta(s, \alpha_1, \mathfrak{B}_1), \ldots, \zeta(s, \alpha_r, \mathfrak{B}_r)$ in the region $D(1/2, 1)$.

In the case $\mathfrak{B}_j = \{1\}_{n=0}^{\infty}$ for $1 \leq j \leq r$ (that is, the case of Hurwitz zeta-functions), the above result was already given in Laurinčikas [91].

In [90] [125], a more general joint strong universality for $\zeta(s, \alpha_j, \mathfrak{B}_{jl})$ $(1 \leq j \leq r,\ 1 \leq l \leq l_j$ with $l_j \in \mathbb{N})$ was discussed under certain matrix conditions.

[§] In [110], the theorem is stated under the weaker assumption that $\alpha_1, \ldots, \alpha_r$ are transcendental, but (as is pointed out in [114]) this assumption should be replaced by the algebraic independence over $\mathbb{Q}$.

Laurinčikas [84] [85] and [38] (with Genys) studied the joint strong universality for general Dirichlet series, under the same assumptions as in Theorem 5.3 and a certain matrix condition.

Now return to the problem of the joint universality for Lerch zeta-functions. Theorem 6.3 implies, especially, that the assumptions of Theorem 6.1 can now be replaced by just the linear independence of (6.1).

Is the assumption (6.1) indeed weaker than the assumptions of Theorem 6.1? The answer is yes, and the following result of Mishou [152] gives an example: Let α_1, α_2 be two transcendental numbers, $0 < \alpha_1, \alpha_2 < 1$, $\alpha_1 \neq \alpha_2$, and $\alpha_2 \in \mathbb{Q}(\alpha_1)$. Then Mishou [152] proved that the joint strong universality holds for $\zeta(s, \alpha_1)$ and $\zeta(s, \alpha_2)$. Dubickas [24] extended Mishou's result to the case of r transcendental numbers, which is also an extension of [91].

Let m_1, m_2, m_3 be relatively prime positive integers (≥ 2), and $\lambda_0 = n_3/m_3$ (with another integer n_3). Nakamura [167] proved the joint strong universality for

$$\left\{\zeta\left(s, \alpha, \lambda_0 + \frac{n_1}{m_1} + \frac{n_2}{m_2}\right) \;\middle|\; 0 \leq n_1 < m_1, 0 \leq n_2 < m_2\right\} \tag{6.2}$$

when α is transcendental. He pointed out that various other types of joint universality can be deduced from the above.

In [115] and in Laurinčikas [92], the joint universality of $\zeta(s, \alpha_j, \lambda_{j\mu_j})$ ($1 \leq j \leq r$, $1 \leq \mu_j \leq m_j$, where m_j is some positive integer) is discussed. Write $\lambda_{j\mu_j} = a_{j\mu_j}/q_{j\mu_j}$, $(a_{j\mu_j}, q_{j\mu_j}) = 1$, and let k_j be the least common multiple of $q_{j\mu_j}$ ($1 \leq \mu_j \leq m_j$). Define $L_j = (\exp(2l\pi i\lambda_{j\mu_j})_{1\leq l\leq k_j, 1\leq \mu_j\leq m_j})$. Then in [92] it is shown that if the elements of the set (6.1) are linearly independent over $\mathbb{Q}$, and $\mathrm{rank}(L_j) = k_j$ ($1 \leq j \leq r$), then the joint strong universality holds for $\zeta(s, \alpha_j, \lambda_{j\mu_j})$.

How about the joint universality for Lerch zeta-functions when the parameter λ is not rational? Nakamura [167] noted that the joint strong universality for (6.2) also holds if we replace λ_0 by any non-rational real number. Also, Nakamura [168] extended the idea in [114] to obtain the following more general result.

Theorem 6.4. *(Nakamura [168]) If $\alpha_1, \ldots, \alpha_r$ are algebraically independent over $\mathbb{Q}$, then for any real numbers $\lambda_1, \ldots, \lambda_r$ the joint strong universality holds for $\zeta(s, \alpha_1, \lambda_1), \ldots, \zeta(s, \alpha_r, \lambda_r)$ in the region $D(1/2, 1)$.*

On the other hand, Mishou [155] proved the following measure-theoretic result.

Theorem 6.5. *(Mishou [155]) There exists a subset $\Lambda \subset [0,1)^r$ whose r-dimensional Lebesgue measure is 1, and for any transcendental real number α and $(\lambda_1, \ldots, \lambda_r) \in \Lambda$, the joint strong universality holds for $\zeta(s, \alpha, \lambda_1), \ldots, \zeta(s, \alpha, \lambda_r)$ in the region $D(1/2, 1)$.*

Moreover in the same paper Mishou gives the following two explicit descriptions of $\lambda_1, \ldots, \lambda_r$ $(0 \le \lambda_j < 1)$ for which the above joint universality holds;

(i) $\lambda_1, \ldots, \lambda_r$ are algebraic irrational and $1, \lambda_1, \ldots, \lambda_r$ are linearly independent over $\mathbb{Q}$,

(ii) $\lambda_1 = \exp(u_1), \ldots, \lambda_r = \exp(u_r)$ where $u_1, \ldots, u_r$ are distinct rational numbers.

Mishou's proof is based on two classical discrepancy estimates due to W. M. Schmidt and H. Niederreiter. These results lead Mishou to propose the following conjecture.

Conjecture 5. *(Mishou [155]) The joint strong universality holds for $\zeta(s, \alpha, \lambda_1), \ldots, \zeta(s, \alpha, \lambda_r)$ in the region $D(1/2, 1)$, for any transcendental real number α $(0 < \alpha < 1)$ and any distinct real numbers $\lambda_1, \ldots, \lambda_r$ $(0 \le \lambda_j < 1)$.*

For single zeta or L-functions, there is Conjecture 4, which asserts that universality would hold for any "reasonable" Dirichlet series. As for the joint universality, the situation is much more complicated. If there is some relation among several Dirichlet series, then the behaviour of those Dirichlet series cannot be independent of each other, so the joint universality among them cannot be expected. Nakamura [168] pointed out that some collections of Lerch zeta-functions cannot be jointly universal, because of the inversion formula among Lerch zeta-functions. In the same paper Nakamura introduced the generalized Lerch zeta-function of the form

$$\zeta(s, \alpha, \beta, \gamma, \lambda) = \sum_{n=0}^{\infty} \frac{e^{2\pi i \lambda n}}{(n+\alpha)^{s-\gamma}(n+\beta)^{\gamma}}, \tag{6.3}$$

and showed that, under suitable choices of parameters the joint strong universality sometimes holds, and sometimes does not hold.

In [169], Nakamura considered more general series

$$\zeta(s, \alpha, \mathcal{C}) = \sum_{n=0}^{\infty} \frac{c(n)}{(n+\alpha)^s}, \tag{6.4}$$

where $\mathcal{C} = \{c(n)\}_{n=0}^{\infty}$ is a bounded sequence of complex numbers, and using it, constructed some counter-examples to the joint universality. The proof in [169] is based on a non-denseness property and a limit theorem on a certain function space.

The above results of Nakamura suggest that it is not easy to find a suitable joint version of Conjecture 4.

Remark 6.1. It is to be noted that there is the following simple principle of producing the joint universality. Let $\varphi \in H(D(\sigma_1, \sigma_2))$, and assume that φ is universal. Let K be a compact subset of $D(\sigma_1, \sigma_2)$ with connected complement, $\lambda_1, \ldots, \lambda_r$ be complex numbers, and $K_j = \{s + \lambda_j \mid s \in K\}$ $(1 \leq j \leq r)$. Assume these K_j's are disjoint. Then $\varphi_j(s) = \varphi(s + \lambda_j)$ $(1 \leq j \leq r)$ are jointly universal. If φ is strongly universal, then φ_j's are jointly strongly universal. This is the *shifts universality principle* of Kaczorowski, Laurinčikas and J. Steuding [62].

7. The universality for multiple zeta-functions

An important generalization of the notion of zeta-functions is multiple zeta-functions, defined by certain multiple sums. The history of the theory of multiple zeta-functions goes back to the days of Euler, but extensive studies started only in 1990s.

The problem of searching for universality theorems on multiple zeta-functions was first proposed by the author [135]. In this paper the author wrote that one accessible problem would be the universality for Barnes multiple zeta-functions

$$\sum_{n_1=0}^{\infty} \cdots \sum_{n_r=0}^{\infty} (w_1 n_1 + \cdots + w_r n_r + \alpha)^{-s}$$

(where $w_1, \ldots, w_r, \alpha$ are parameters). Nakamura [168] pointed out that his $\zeta(s, \alpha, \beta, \gamma, \lambda)$ (see (6.3)) includes the twisted Barnes double zeta-function of the form

$$\sum_{n_1=0}^{\infty} \sum_{n_2=0}^{\infty} \frac{e^{2\pi i \lambda (n_1 + n_2)}}{(n_1 + n_2 + \alpha)^s}$$

as a special case. Therefore [168] includes a study of universality for Barnes double zeta-functions. More generally, the series $\zeta(s, \alpha, \mathcal{C})$ (see (6.4)) studied in Nakamura [169] includes twisted Barnes r-ple zeta-functions (for any r).

The Euler-Zagier r-ple sum is defined by

$$\sum_{n_1 > n_2 > \cdots n_r \geq 1} n_1^{-s_1} n_2^{-s_2} \cdots n_r^{-s_r}, \tag{7.1}$$

where $s_1,\ldots,s_r$ are complex variables. Nakamura [170] considered the universality of the following generalization of (7.1) of Hurwitz-type:

$$\zeta_r(s_1,\ldots,s_r;\alpha_1,\ldots,\alpha_r)= \sum_{n_1>n_2>\cdots n_r\geq 0}(n_1+\alpha_1)^{-s_1}(n_2+\alpha_2)^{-s_2}\cdots(n_r+\alpha_r)^{-s_r}, \tag{7.2}$$

where $0<\alpha_j\leq 1$ $(1\leq j\leq r)$. Nakamura's results suggest that universality for the multiple zeta-function (7.2) is connected with the zero-free region. One of his main results is as follows.

Theorem 7.1. *(Nakamura [170]) Let $\Re s_2>3/2$, $\Re s_j\geq 1$ $(3\leq j\leq r)$. Assume α_1 is transcendental, and $\zeta_{r-1}(s_2,\ldots,s_r;\alpha_2,\ldots,\alpha_r)\neq 0$. Then the strong universality holds for $\zeta_r(s_1,\ldots,s_r;\alpha_1,\ldots,\alpha_r)$ as a function in s_1 in the region $D(1/2,1)$.*

In [173], Nakamura considered a generalization of Tornheim's double sum of Hurwitz-type and proved the strong universality for it.

8. The mixed universality

In Section 4 we discussed the joint universality among zeta or L-functions with Euler products. Then in Section 6 we considered the joint universality for those without Euler products. Is it possible to combine these two directions to obtain certain joint universality results between two (or more) zeta-functions, one of which has Euler products and the other does not? The first affirmative answers are due to Sander and J. Steuding [191], and to Mishou [149]. The work of Sander and J. Steuding will be discussed later in Section 15. Here we state Mishou's theorem.

Theorem 8.1. *(Mishou [149]) Let K_1,K_2 be compact subsets of $D(1/2,1)$ with connected complements, and $f_1\in H_0^c(K_1)$, $f_2\in H^c(K_2)$. Then, for any $\varepsilon>0$, we have*

$$\liminf_{T\to\infty}\frac{1}{T}\mathrm{meas}\left\{\tau\in[0,T]\;\middle|\;\sup_{s\in K_1}|\zeta(s+i\tau)-f_1(s)|<\varepsilon,\right. \tag{8.1}$$
$$\left.\sup_{s\in K_2}|\zeta(s+i\tau,\alpha)-f_2(s)|<\varepsilon\right\}>0,$$

provided α is transcendental.

This type of universality is now called the *mixed universality*. The essential point of the proof of this theorem is the fact that the elements of the set

$$\{\log(n+\alpha) \mid n \in \mathbb{N}_0\} \cup \{\log p \mid p : \text{prime}\}$$

are linearly independent over $\mathbb{Q}$.

Mishou's theorem was generalized to the periodic case by Kačinskaitė and Laurinčikas [58]¶. Let $\mathfrak{A}$ be a multiplicative periodic sequence satisfying (3.1) and $\mathfrak{B}$ a (not necessarily multiplicative) periodic sequence. They proved that if α is transcendental, then the mixed universality holds for $\zeta(s,\mathfrak{A})$ and $\zeta(s,\alpha,\mathfrak{B})$ in the region $D(1/2,1)$.

Laurinčikas [93] proved a further generalization. Let $\mathfrak{A}_1,\ldots,\mathfrak{A}_{r_1}$ be multiplicative periodic sequences (with inequalities similar to (3.1)) and $\mathfrak{B}_1,\ldots,\mathfrak{B}_{r_2}$ be periodic sequences. Then, under certain matrix conditions, the mixed universality for $\zeta(s,\mathfrak{A}_{j_1})$ $(1 \le j_1 \le r_1)$ and $\zeta(s,\alpha_{j_2},\mathfrak{B}_{j_2})$ $(1 \le j_2 \le r_2)$ holds, provided $\alpha_1,\ldots,\alpha_{r_2}$ are algebraically independent over $\mathbb{Q}$.

Now mixed universality theorems are known for many pairs of zeta or L-functions.

- The Riemann zeta-function and several periodic Hurwitz zeta-functions (Genys, Macaitienė, Račkauskienė and Šiaučiūnas [39]),
- The Riemann zeta-function and several Lerch zeta-functions (Laurinčikas and Macaitienė [106]),
- Several Dirichlet L-functions and several Hurwitz zeta-functions (Janulis and Laurinčikas [47]),
- Several Dirichlet L-functions and several periodic Hurwitz zeta-functions (Janulis, Laurinčikas, Macaitienė and Šiaučiūnas [48]),
- An automorphic L-function and several periodic Hurwitz zeta-functions (Laurinčikas, Macaitienė and Šiaučiūnas [109], Macaitienė [131], Pocevičienė and Šiaučiūnas [186], Laurinčikas and Šiaučiūnas [123]).

9. The strong recurrence

In Section 5 we mentioned that the strong universality implies the existence of many zeros in the region where universality is valid (Theorem 5.2). This immediately gives the following corollary:

Corollary 9.1. *The Riemann zeta-function $\zeta(s)$ cannot be strongly universal in the region $D(1/2,1)$.*

¶This paper was published in 2011, but was already completed in 2007.

Because if $\zeta(s)$ is strongly universal, then by Theorem 5.2 we have $N(T;\sigma_1,\sigma_2;\zeta) \geq CT$ for $1/2 < \sigma_1 < \sigma_2 < 1$, which contradicts with the known zero-density estimate $N(T;\sigma_1,\sigma_2;\zeta) = o(T)$.

The same conclusion can be shown for many other zeta or L-functions, for which some suitable zero-density estimate is known; or, under the assumption of the analogue of the Riemann hypothesis.

On the other hand, if the Riemann hypothesis is true, then $\zeta(s)$ has no zero in the region $D(1/2,1)$. Therefore we can choose $f(s) = \zeta(s)$ in Theorem 1.1 to obtain

$$\liminf_{T\to\infty} \frac{1}{T}\text{meas}\left\{\tau \in [0,T] \;\middle|\; \sup_{s\in K} |\zeta(s+i\tau) - \zeta(s)| < \varepsilon \right\} > 0. \tag{9.1}$$

This is called the *strong recurrence* property of $\zeta(s)$. Bagchi discovered that the converse implication is also true.

Theorem 9.1. *(Bagchi [3]) The Riemann hypothesis for $\zeta(s)$ is true if and only if* (9.1) *holds in the region $D(1/2,1)$.*

Bagchi himself extended this result to the case of Dirichlet L-functions in [4] [5]. The same type of result in terms of Beurling zeta-functions was given by R. Steuding [203].

Remark 9.1. It is obvious that the notion of the strong recurrence is closely connected with the notion of almost periodicity. Bohr [12] proved that the Riemann hypothesis for $L(s,\chi)$ with a non-principal character χ is equivalent to the almost periodicity of $L(s,\chi)$ in the region $\Re s > 1/2$. Recently Mauclaire [138] [139] studied the universality in a general framework from the viewpoint of almost periodicity.

Nakamura [171] proved that, if $d_1 = 1, d_2, \ldots, d_r$ are algebraic real numbers which are linearly independent over $\mathbb{Q}$, then the joint universality of the form

$$\liminf_{T\to\infty} \frac{1}{T}\text{meas}\left\{\tau \in [0,T] \;\middle|\; \sup_{s\in K_j} |\zeta(s+id_j\tau) - f_j(s)| < \varepsilon \quad (1 \leq j \leq r)\right\} > 0 \tag{9.2}$$

holds, where K_j are compact subsets of $D(1/2,1)$ with connected complements and $f_j \in H_0^c(K_j)$. A key point of Nakamura's proof is the fact that the elements of the set $\{\log p^{d_j} \mid p : \text{prime}, 1 \leq j \leq r\}$ are linearly independent over $\mathbb{Q}$, which follows from Baker's theorem in transcendental number theory. From the above result it is immediate that

$$\liminf_{T\to\infty} \frac{1}{T}\text{meas}\left\{\tau \in [0,T] \;\middle|\; \sup_{s\in K} |\zeta(s+i\tau) - \zeta(s+id\tau)| < \varepsilon\right\} > 0 \quad (9.3)$$

holds if d is algebraic irrational. Nakamura also proved in the same paper that (9.3) is valid for almost all $d \in \mathbb{R}$. (Note that (9.3) for $d = 0$ is (9.1), hence the Riemann hypothesis.)

Nakamura's paper sparked off the interest in this direction of research; Pańkowski [182] proved that (9.3) holds for all (algebraic and transcendental) irrational d, using the six exponentials theorem in transcendental number theory. On the other hand, Garunkštis [34] and Nakamura [172], independently, claimed that (9.3) holds for all non-zero rational. However their arguments included a gap, which was partially filled by Nakamura and Pańkowski [176]. The present situation is:

Theorem 9.2. *(Garunkštis, Nakamura, Pańkowski) The inequality* (9.3) *holds if d is irrational, or $d = a/b$ is non-zero rational with* $(a,b) = 1$, $|a-b| \neq 1$.

See also Mauclaire [140], and Nakamura and Pańkowski [178]. It is to be noted that the argument of Garunkštis [34] and Nakamura [172] is correct for $\log\zeta(s)$, so

$$\liminf_{T\to\infty} \frac{1}{T}\text{meas}\left\{\tau \in [0,T] \;\middle|\; \sup_{s\in K} |\log\zeta(s+i\tau) - \log\zeta(s+id\tau)| < \varepsilon\right\} > 0 \quad (9.4)$$

can be shown for any non-zero $d \in \mathbb{R}$. If (9.4) would be valid for $d = 0$, it would imply the Riemann hypothesis.

The strong recurrence property can be shown for more general zeta and L-functions. Some of the aforementioned papers actually consider not only the Riemann zeta-function, but also Dirichlet L-functions. A generalization to a subclass of the Selberg class was discussed by Nakamura [174]. The case of Hurwitz zeta-functions has been studied by Garunkštis and Karikovas [35] and Karikovas and Pańkowski [65].

10. The weighted universality

Laurinčikas [75] considered a weighted version of the universality for $\zeta(s)$. Let $w(t)$ be a positive-valued function of bounded variation defined on $[T_0, \infty)$ (where $T_0 > 0$), satisfying that the variation on $[a, b]$ does not exceed $cw(a)$ with a certain $c > 0$ for any subinterval $[a, b] \subset [T_0, \infty)$. Define

$$U(T, w) = \int_{T_0}^{T} w(t)dt,$$

and assume that $U(T, w) \to \infty$ as $T \to \infty$.

We further assume the following property of $w(t)$, connected with ergodic theory. Let $X(\tau, \omega)$ be any ergodic process defined on a certain probability space Ω, $\tau \in \mathbb{R}$, $\omega \in \Omega$, $E(|X(\tau, \omega)|) < \infty$, and sample paths are Riemann integrable almost surely on any finite interval. Assume that

$$\frac{1}{U(T, w)} \int_{T_0}^{T} w(\tau) X(t + \tau, \omega) d\tau = E(X(0, \omega)) + o((1 + |t|)^{\alpha}) \tag{10.1}$$

almost surely for any $t \in \mathbb{R}$, with an $\alpha > 0$, as $T \to \infty$. Denote by $I(A)$ the indicator function of the set A.

Theorem 10.1. *(Laurinčikas [75]) Suppose that $w(t)$ satisfies all the above conditions. Let K be a compact subset of $D(1/2, 1)$ with connected complement, $f \in H_0^c(K)$. Then*

$$\liminf_{T \to \infty} \frac{1}{U(T, w)} \int_{T_0}^{T} w(\tau) \times I\left(\left\{\tau \in [T_0, T] \;\middle|\; \sup_{s \in K} |\zeta(s + i\tau) - f(s)| < \varepsilon\right\}\right) d\tau > 0 \tag{10.2}$$

holds for any $\varepsilon > 0$.

In the course of Bagchi's proof of the universality theorem, there is a point where the Birkhoff-Khinchin theorem

$$\lim_{T \to \infty} \frac{1}{T} \int_{0}^{T} X(\tau, \omega) d\tau = E(X(0, \omega)) \tag{10.3}$$

in ergodic theory is used. This is the motivation of Laurinčikas [75]; clearly (10.1) is a generalization of (10.3).

Laurinčikas [78] generalized Theorem 10.1 to the case of Matsumoto zeta-functions. Weighted universality theorems for L-functions $L(s, E)$ of elliptic curves over $\mathbb{Q}$ were reported by Garbaliauskienė [26] [27].

11. The discrete universality

In the previous sections, we discussed the behaviour of zeta or L-functions when the imaginary part τ of the variable is moving continuously. However, we can also obtain a kind of universality theorems when τ only moves discretely. We already mentioned in Section 1 that Voronin's multi-dimensional denseness theorem is valid in this sense (see Remark 1.3).

The first *discrete universality* theorem is due to Reich [189] on Dedekind zeta-functions. Let F be a number field and $\zeta_F(s)$ the associated Dedekind zeta-function.

Theorem 11.1. *(Reich [189]) Let K be a compact subset of the region $D(1-\max\{2,d_F\}^{-1},1)$ with connected complement, and $f \in H_0^c(K)$. Then, for any real $h \neq 0$ and any $\varepsilon > 0$,*

$$\liminf_{N\to\infty} \frac{1}{N}\#\left\{n \le N \;\middle|\; \sup_{s\in K} |\zeta_F(s+ihn) - f(s)| < \varepsilon\right\} > 0. \tag{11.1}$$

The joint discrete universality theorem for Dirichlet L-functions was given in Bagchi [3]. He also obtained the discrete universality for $\zeta(s,\alpha)$ when α is rational, for which Sander and J. Steuding [191] gave a different approach.

Kačinskaitė [54] (see also [56]) proved a discrete universality theorem for Matsumoto zeta-functions under the condition that $\exp(2\pi k/h)$ is irrational for any non-zero integer k. Ignatavičiūtė [46] reported that certain discrete universality and certain joint discrete universality hold for Lerch zeta-functions, provided $\exp(2\pi/h)$ is rational.

As can be seen in the above examples, an interesting point on discrete universality is that the arithmetic nature of the parameter h plays a role.

The discrete universality for L-functions $L(s,E)$ of elliptic curves was first studied in [30] under the assumption that $\exp(2\pi k/h)$ is irrational for any non-zero integer k. However this condition was then removed:

Theorem 11.2. *(Garbaliauskienė, Genys and Laurinčikas [28]) The discrete universality holds for $L(s,E)$ for any real $h \neq 0$ in the region $D(1,3/2)$.*

The same type of result can be shown, more generally, for L-functions attached to new forms. When $\exp(2\pi k/h)$ is irrational for any non-zero integer k, this was done in Laurinčikas, Matsumoto and J. Steuding [117].

The discrete universality for periodic zeta-functions was studied in Kačinskaitė, Javtokas and Šiaučiūnas [57] and Laurinčikas, Macaitienė and Šiaučiūnas [108], while the case of periodic Hurwitz zeta-functions was discussed by Laurinčikas and Macaitienė [104]. The result in [104] especially includes the discrete universality of the Hurwitz zeta-function $\zeta(s,\alpha)$ when

α is transcendental. Laurinčikas [100] further proved that if the set

$$\{\log(m+\alpha) \mid m \in \mathbb{N}_0\} \cup \{2\pi/h\}$$

is linearly independent over $\mathbb{Q}$, then the discrete universality holds for $\zeta(s,\alpha)$. See also [18]. A joint version is studied in Laurinčikas [101].

Macaitienė [130] obtained a discrete universality theorem for general Dirichlet series.

Theorem 11.3. *(Macaitienė [130]) Let $f(s)$ be general Dirichlet series as in Theorem 5.3, and further suppose that λ_n are algebraic numbers and* $\exp(2\pi/h) \in \mathbb{Q}$. *Then the discrete universality holds for $f(s)$ in the region $D(\sigma_1, \sigma_a)$.*

The discrete analogue of mixed universality can also be considered. This direction was first studied by Kačinskaitė [55]. Consider the case $\exp(2\pi/h) \in \mathbb{Q}$. Write $\exp(2\pi/h) = a/b$ with $a, b \in \mathbb{Z}$ and $(a,b) = 1$. Denote by P_h the set of all prime numbers appearing as a prime factor of a or b. Define the modified Dirichlet L-function $L_h(s,\chi)$ by removing all Euler factors corresponding to primes in P_h, that is

$$L_h(s,\chi) = \prod_{p \notin P_h} \left(1 - \frac{\chi(p)}{p^s}\right)^{-1}.$$

Theorem 11.4. *(Kačinskaitė [55]) Let K_1, K_2 be compact subsets of $D(1/2,1)$ with connected complements, and $f_1 \in H_0^c(K_1)$, $f_2 \in H^c(K_2)$. If α is transcendental, $\mathfrak{B}$ is a periodic sequence and* $\exp(2\pi/h) \in \mathbb{Q}$ *as above, then*

$$\liminf_{N\to\infty} \frac{1}{N} \# \left\{ n \le N \;\middle|\; \sup_{s\in K_1} |L_h(s+ihn,\chi) - f_1(s)| < \varepsilon, \right. \tag{11.2}$$
$$\left. \sup_{s\in K_2} |\zeta(s+ihn,\alpha,\mathfrak{B}) - f_2(s)| < \varepsilon \right\} > 0$$

for any $\varepsilon > 0$. [‖]

Buivytas and Laurinčikas [16] proved that if the set

$$\{\log p \mid p : \text{prime}\} \cup \{\log(m+\alpha) \mid m \in \mathbb{N}_0\} \cup \{2\pi/h\}$$

[‖]The statement in [55] is given for $L(s,\chi)$, but her argument is valid not for $L(s,\chi)$, but for $L_h(s,\chi)$.

is linearly independent over $\mathbb{Q}$, then the discrete mixed universality holds for $\zeta(s)$ and $\zeta(s,\alpha)$, that is,

$$\liminf_{N\to\infty}\frac{1}{N}\#\left\{n\le N \;\middle|\; \sup_{s\in K_1}|\zeta(s+ihn)-f_1(s)|<\varepsilon, \right. \tag{11.3}$$
$$\left. \sup_{s\in K_2}|\zeta(s+ihn,\alpha)-f_2(s)|<\varepsilon\right\}>0.$$

In (11.2) and (11.3), the shifting parameter h is common to the both of relevant zeta (or L)-functions. Buivytas and Laurinčikas [17] studied the case when the parameter for $\zeta(s)$ is different from the parameter for $\zeta(s,\alpha)$.

We will encounter a rather different type of discrete universality theorems in Section 18.

12. The χ-universality

The main point of Voronin's universality theorem is the existence of τ, the imaginary part of the complex variable s, satisfying a certain approximation condition. This is the common feature of all universality theorems mentioned in the previous sections. However there is another type of universality, which is concerned with the existence of a character satisfying certain approximation conditions.

The first theorem in this direction is as follows.

Theorem 12.1. *(Bagchi [3], Gonek [40], Eminyan [25]) Let K be a compact subset of $D(1/2,1)$ with connected complement, and $f\in H_0^c(K)$. Let Q be an infinite set of positive integers. Then for any $\varepsilon>0$,*

$$\liminf_{\substack{q\to\infty\\ q\in Q}}\frac{1}{\varphi(q)}\#\left\{\chi(\mathrm{mod}\ q)\;\middle|\;\sup_{s\in K}|L(s,\chi)-f(s)|<\varepsilon\right\}>0 \tag{12.1}$$

*holds, provided Q is one of the following**:*

(i) Q is the set of all prime numbers;

(ii) Q is the set of positive integers of the form $q=p_1^{a_1}\cdots p_r^{a_r}$ $(a_1,\ldots,a_r\in\mathbb{N}\cup\{0\})$, where $\{p_1,\ldots,p_r\}$ is a fixed finite set of prime numbers.

This type of results is called the *χ-universality*, or *the universality in χ-aspect*. The universality for Hecke L-functions of number fields in χ-aspect was discussed by Mishou and Koyama [157], and by Mishou [147] [148].

**Bagchi [3, Theorem 5.3.11] proved the case (i). In the statement of Gonek [40, Theorem 5.1] there is no restriction on Q, but Mishou [146] pointed out that condition (ii) is necessary to verify Gonek's proof. Eminyan [25] studied the special case $r=1$ of (ii).

Let χ_d be a real Dirichlet character with discriminant d. Another interesting direction of research is the universality for $L(s, \chi_d)$ in d-aspect. This direction was studied in a series of papers by Mishou and Nagoshi [158] [159] [160] [161]. Let Λ^+ (resp. Λ^-) be the set of all positive (resp. negative) discriminants, and $\Lambda^+(X)$ (resp. $\Lambda^-(X)$) be the set of discriminants d satisfying $0 < d \leq X$ (resp. $-X \leq d < 0$).

Theorem 12.2. *(Mishou and Nagoshi [158]) Let Ω be a simply connected domain in $D(1/2, 1)$ which is symmetric with respect to the real axis. Let $f(s)$ be holomorphic and non-vanishing on Ω, and positive-valued on $\Omega \cap \mathbb{R}$. Let K be a compact subset of Ω. Then, for any $\varepsilon > 0$,*

$$\liminf_{X\to\infty} \frac{1}{\#\Lambda^\pm(X)} \# \left\{ d \in \Lambda^\pm(X) \;\middle|\; \sup_{s\in K} |L(s,\chi_d) - f(s)| < \varepsilon \right\} > 0 \quad (12.2)$$

holds.

This theorem especially implies that for any $s \in D(1/2, 1) \setminus \mathbb{R}$, the set $\{L(s, \chi_d) \mid d \in \Lambda^\pm\}$ is dense in $\mathbb{C}$, and for any real number σ with $1/2 < \sigma < 1$, the set $\{L(\sigma, \chi_d) \mid d \in \Lambda^\pm\}$ is dense in the set of positive real numbers $\mathbb{R}_+$.

In the same paper Mishou and Nagoshi also studied the situation on the line $\Re s = 1$, and proved that the set $\{L(1, \chi_d) \mid d \in \Lambda^\pm\}$ is dense in $\mathbb{R}_+$. Therefore we can deduce denseness results on class numbers of quadratic fields. Let $h(d)$ be the class number of $\mathbb{Q}(\sqrt{d})$, and when $d > 0$, let $\varepsilon(d)$ be the fundamental unit of $\mathbb{Q}(\sqrt{d})$. Then the above result implies that both the sets

$$\left\{ \frac{h(d) \log \varepsilon(d)}{\sqrt{d}} \;\middle|\; d \in \Lambda^+ \right\}, \qquad \left\{ \frac{h(d)}{\sqrt{d}} \;\middle|\; d \in \Lambda^- \right\}$$

are dense in $\mathbb{R}_+$.

In [159] [161], the same type of problem for prime discriminants are studied. As an application, in [161] Mishou and Nagoshi gave a quantitative result on a problem of Ayoub, Chowla and Walum on certain character sums [2]. In [160] Mishou and Nagoshi gave some conditions equivalent to the Riemann hypothesis from their viewpoint.

The universality in d-aspect for Hecke L-functions of class group characters for imaginary quadratic fields are studied by Mishou [151]. In [150], Mishou considered cubic characters associated with the field $\mathbb{Q}(\sqrt{-3})$, and proved a universality theorem for associated Hecke L-functions in cubic character aspect.

13. Other applications

So far we mentioned applications of universality to the theory of distribution of zeros (Section 5), to the Riemann hypothesis (Section 9), and to algebraic number theory (Section 12). Those applications, however, do not exhaust the potentiality of universality theory. In this section we discuss other applications of universality.

In Section 1 we mentioned that Voronin, before proving his universality theorem, obtained the multi-dimensional denseness theorem (Theorem 1.2) of $\zeta(s)$ and its derivatives. However, now, we can say that Theorem 1.2 is just an immediate consequence of the universality theorem (see [76, Theorem 6.6.2]). Moreover, from Theorem 1.2 it is easy to obtain the following functional-independence property of $\zeta(s)$.

Theorem 13.1. *Let $f_l : \mathbb{C}^m \to \mathbb{C}$ $(0 \leq l \leq n)$ be continuous functions, and assume that the equality*

$$\sum_{l=0}^{n} s^l f_l(\zeta(s), \zeta'(s), \ldots, \zeta^{(m-1)}(s)) = 0 \tag{13.1}$$

holds for all s. Then $f_l \equiv 0$ for $0 \leq l \leq n$.

This type of result was first noticed by Voronin himself ([206] [209]); see also [76, Theorem 6.6.3].

When f_l's are polynomials, then (13.1) is an algebraic differential equation. Therefore Theorem 13.1 in this case implies that $\zeta(s)$ does not satisfy any non-trivial algebraic differential equation. This property was already noticed by Hilbert in his famous address [45] at the 2nd International Congress of Mathematicians (Paris, 1900). Theorem 13.1 is a generalization of this algebraic-independence property.

Similarly to the case of $\zeta(s)$, if a Dirichlet series $\varphi(s)$ is universal, it is easy to prove the theorems analogous to Theorem 1.2 and Theorem 13.1 for $\varphi(s)$.

An application of the universality to the problem on Dirichlet polynomials was done by Andersson [1]. He used the universality theorem to show that several conjectures proposed by Ramachandra [187] and Balasubramanian and Ramachandra [6], on lower bounds of certain integrals of Dirichlet polynomials, are false.

The universality property was applied even in physics; see Gutzwiller [44], Bitar, Khuri and Ren [10].

14. The general notion of universality

The main theme of the present article is the universality for zeta and L-functions. However, the notion of universality was first introduced in mathematics, under a very different motivation.

The first discovery of the universality phenomenon is usually attributed to M. Fekete (1914/15, reported in [181]), who proved that there exists a real power series $\sum_{n=1}^{\infty} a_n x^n$ such that, for any continuous $f : [-1,1] \to \mathbb{R}$ with $f(0) = 0$ we can choose positive integers $m_1, m_2, \ldots$ for which

$$\sum_{n=1}^{m_k} a_n x^n \to f(x) \qquad (k \to \infty)$$

holds uniformly on $[-1,1]$. The proof is based on Weierstrass' approximation theorem.

G. D. Birkhoff [9] proved that there exists an entire function $\psi(z)$ such that, for any entire function $f(z)$, we can choose complex numbers $a_1, a_2, \ldots$ for which $\psi(z + a_k) \to f(z)$ (as $k \to \infty$) uniformly in any compact subset of $\mathbb{C}$.

The terminology "universality" was first used by Marcinkiewicz [132]. Various functions satisfying some property similar to those discovered by Fekete and Birkhoff are known. However, before the work of Voronin [205], all of those functions were constructed very artificially. So far the class of zeta and L-functions is the only "natural" class of functions for which the universality property can be proved. For the more detailed history of this general notion of universality, see Grosse-Erdmann [43] and Steuding [201, Appendix]. [††]

It is to be noted that the real origin of the whole theory is Riemann's theorem that a conditionally convergent series can be convergent (or divergent) to any value after some suitable rearrangement. In fact, Fekete's result may be regarded as an analogue of Riemann's theorem for continuous functions, while Pecherskiĭ's theorem [185] (mentioned in Section 1 as an essential tool in Voronin's proof) gives an analogue of Riemann's theorem in Hilbert spaces.

A very general definition of universality was proposed by Grosse-Erdmann [42] [43].

Definition 14.1. Let X, Y be topological spaces, W be a non-empty closed subset of Y, and $T_\tau : X \to Y$ ($\tau \in I$) be a family of mappings with the

[††] In the same Appendix, Steuding mentioned a p-adic version of Fekete's theorem, which was originally proved in [196].

index set I. We call $x \in X$ *universal with respect to* W if the closure of the set $\{T_\tau(x) \mid \tau \in I\}$ contains W.

Let K be as in Theorem 1.1, $X = Y = H(K^\circ)$, where K° is the interior of K. Then obviously $H_0^c(K) \subset H(K^\circ)$. Put $W = \overline{H_0^c(K)}$ (the topological closure of $H_0^c(K)$ in the space $H(K^\circ)$). Define T_τ by $T_\tau(f(z)) = f(z + i\tau)$ for $f \in H(K^\circ)$. Then Theorem 1.1 implies that any element of $H_0^c(K)$ can be approximated by some suitable element of $\{T_\tau(\zeta) \mid \tau \in \mathbb{R}\}$. Therefore Theorem 1.1 asserts that the Riemann zeta-function $\zeta(s)$ is universal with respect to $\overline{H_0^c(K)}$ in the sense of Definition 14.1.

The notion of joint universality can also be formulated in this general setting.

Definition 14.2. Let $X, Y_1, \ldots, Y_r$ be topological spaces, and $T_\tau^{(j)} : X \to Y_j$ $(\tau \in I, 1 \leq j \leq r)$ be families of mappings. We call $x_1, \ldots, x_r \in X$ *jointly universal* if the set $\{(T_\tau^{(1)}(x_1), \ldots, T_\tau^{(r)}(x_r) \mid \tau \in I\}$ is dense in $Y_1 \times \cdots \times Y_r$.

Remark 14.1. The case $r = 1$ of Definition 14.2 is the case $W = Y$ of Definition 14.1.

In Section 1 we mentioned that the Kronecker-Weyl approximation theorem (see Remark 1.1) is used in the proof of Theorem 1.1. We can see that the Kronecker-Weyl theorem itself implies a certain universality phenomenon. Let $S^1 = \{z \in \mathbb{C} \mid |z| = 1\}$, and consider the situation when $X = \mathbb{R}$, $Y_1 = \cdots = Y_r = S^1$ in Definition 14.2. Define $T_\tau^{(1)} = \cdots = T_\tau^{(r)} = T_\tau : \mathbb{R} \to S^1$ by $T_\tau(x) = e^{2\pi i\tau x}$. Then the Kronecker-Weyl theorem implies that, if $\alpha_1, \ldots, \alpha_r \in \mathbb{R}$ are linearly independent over $\mathbb{Q}$, then the orbit of

$$(T_\tau(\alpha_1), \ldots, T_\tau(\alpha_r)) = (e^{2\pi i\tau\alpha_1}, \ldots, e^{2\pi i\tau\alpha_r})$$

is dense in $S^1 \times \cdots \times S^1$. Therefore $\alpha_1, \ldots, \alpha_r$ are jointly universal.

The above observation shows that both the Voronin theorem and the Kronecker-Weyl theorem express certain universality properties. Is it possible to combine these two universality theorems? The answer is yes, and we will discuss this matter in the next section.

15. The hybrid universality

The first affirmative answer to the question raised at the end of Section 14 was given by Gonek [40], and a slightly general result was later obtained by Kaczorowski and Kulas [61].

Theorem 15.1. *(Gonek [40], Kaczorowski and Kulas [61]) Let K be a compact subset of $D(1/2,1)$, $f_1,\ldots,f_r \in H_0^c(K)$, $\chi_1,\ldots,\chi_r$ be pairwise non-equivalent Dirichlet characters, $z>0$, and $(\theta_p)_{p\le z}$ be a sequence of real numbers indexed by prime numbers up to z. Then, for any $\varepsilon>0$,*

$$\liminf_{T\to\infty}\frac{1}{T}\mathrm{meas}\left\{\tau\in[0,T]\;\middle|\;\max_{1\le j\le r}\sup_{s\in K}|L(s+i\tau,\chi_j)-f_j(s)|<\varepsilon,\right. \tag{15.1}$$
$$\left.\max_{p\le z}\left\|\tau\frac{\log p}{2\pi}-\theta_p\right\|<\varepsilon\right\}>0$$

holds.

The combination of the universality of Voronin type and of Kronecker-Weyl type is now called the *hybrid universality*. The above theorem is therefore an example of the hybrid joint universality.

Pańkowski [183] proved that the second inequality in (15.1) can be replaced by $\max_{1\le k\le m}||\tau\alpha_k-\theta_k||<\varepsilon$, where $\alpha_1,\ldots,\alpha_m$ are real numbers which are linearly independent over $\mathbb{Q}$, and $\theta_1,\ldots,\theta_m$ are arbitrary real numbers. This is exactly the same inequality as in the Kronecker-Weyl theorem (see Remark 1.1).

In the same paper Pańkowski remarked that the same statement can be shown for more general L-functions which have Euler products. The hybrid joint universality for some zeta-functions without Euler products was discussed in Pańkowski [184].

Hybrid universality theorems are quite useful in applications. Gonek [40] used his hybrid universality theorem to show the joint universality for Dedekind zeta-functions of Abelian number fields (mentioned in Section 3). The key fact here is that those Dedekind zeta-functions can be written as products of Dirichlet L-functions. On the other hand, the aim of Kaczorowski and Kulas [61] was to study the distribution of zeros of linear combinations of the form $\sum P_j(s)L(s,\chi_j)$, where $P_j(s)$ are Dirichlet polynomials. Kaczorowski and Kulas applied Theorem 15.1 to show a theorem[‡‡], similar to Theorem 5.2, for such linear combinations.

Sander and J. Steuding [191] also considered the universality for sums, or products of Dirichlet L-functions. In particular they proved the joint universality for Hurwitz zeta-functions $\zeta(s,a/q)$ $(1\le a\le q)$ under a certain condition on target functions. Hurwitz zeta-functions usually do not have Euler products, but when $a=q$ (and when q is even and $a=q/2$)

[‡‡]This theorem was later sharpened by Ki and Lee [66] by using the method of mean motions [52] [15].

the corresponding Hurwitz zeta-function is essentially the Riemann zeta-function and hence has the Euler product. Therefore the result of Sander and J. Steuding is an example of mixed universality (see Section 8).

In the paper of Kaczorowski and Kulas [61], the coefficients $P_j(s)$ of linear combinations are Dirichlet polynomials. Nakamura and Pańkowski [177] considered a more general situation when $P_j(s)$ are Dirichlet series. Their general statement is as follows.

Theorem 15.2. *(Nakamura and Pańkowski [177]) Let $P_1(s), \ldots, P_r(s)$ $(r \geq 2)$ be general Dirichlet series, not identically vanishing, absolutely convergent in $\Re s > 1/2$. Moreover assume that at least two of those are non-vanishing in $D(1/2, 1)$. Let $L_1(s), \ldots, L_r(s)$ be hybridly jointly universal in the above sense. Then $L(s) = \sum_{j=1}^{r} P_j(s) L_j(s)$ is strongly universal in $D(1/2, 1)$.*

As a corollary, by Theorem 5.2 we find $N(T; \sigma_1, \sigma_2; L) \geq CT$ for the above $L(s)$, for any σ_1, σ_2 satisfying $1/2 < \sigma_1 < \sigma_2 < 1$.

In the above theorem $L(s)$ is a linear form of L_j's, but in [175] [180], Nakamura and Pańkowski obtained more general statements; they considered polynomials of L_j's whose coefficients are general Dirichlet series, and proved results similar to Theorem 15.2. Note that when coefficients of polynomials are constants, such a result was already given in Kačinskaitė, J. Steuding, Šiaučiūnas and Šleževičienė [59].

Many important zeta and L-functions have such polynomial expressions. Consequently, Nakamura and Pańkowski succeeded in proving the inequalities like (5.3) on the distribution of zeros of those zeta or L-functions, such as zeta-functions attached to symmetric matrices (in the theory of prehomogeneous vector spaces), Estermann zeta-functions, Igusa zeta-functions associated with local Diophantine problems, spectral zeta-functions associated with Laplacians on Riemannian manifolds, Epstein zeta-functions (see [179]), and also various multiple zeta-functions (of Euler-Zagier, of Barnes, of Shintani, of Witten and so on).

16. Quantitative results

It is an important question how to obtain quantitative information related with universality. For example, let

$$d(\zeta, f, K, \varepsilon) = \liminf_{T \to \infty} \frac{1}{T} \text{meas} \left\{ \tau \in [0, T] \;\middle|\; \sup_{s \in K} |\zeta(s + i\tau) - f(s)| < \varepsilon \right\}. \tag{16.1}$$

Theorem 1.1 asserts that $d(\zeta, f, K, \varepsilon)$ is positive; but how to evaluate this value? Or, how to find the smallest value of τ (which we denote by $\tau(\zeta, f, K, \varepsilon)$) satisfying the inequality $\sup_{s\in K} |\zeta(s+i\tau) - f(s)| < \varepsilon$? Voronin's proof gives no information on these questions, because in the course of the proof Voronin used Pecherskiĭ's rearrangement theorem, which is ineffective.

The first attempt to get a quantitative version of the universality theorem is due to Good [41]. The fundamental idea of Good is to combine the argument of Voronin with the method of Montgomery [162], in which Montgomery studied large values of $\log \zeta(s)$. Instead of Pecherskiĭ's theorem, Good used convexity arguments (Hadamard's three circles theorem and the Hahn-Banach theorem). Also Koksma's quantitative version of Weyl's criterion is invoked. The statements of Good's main results are quite complicated, but it includes a quantitative version of the discrete universality theorem for $\zeta(s)$.

Remark 16.1. Actually Good stated the discrete universality for $\log \zeta(s)$. From the universality for $\log \zeta(s)$, the universality for $\zeta(s)$ itself can be immediately deduced by exponentiation. A proof of the universality for $\log \zeta(s)$ by Voronin's original method is presented in the book of Karatsuba and Voronin [64, Chapter VII].

Good's idea was further pursued by Garunkštis [33], who obtained a more explicit quantitative result when K is small. His main theorem is still rather complicated, but as a typical special case, he showed the following inequalities. Let $K = K(r)$ be as in Section 1.

Theorem 16.1. *(Garunkštis [33]) Let* $0 < \varepsilon \le 1/2$, $f \in H(K(0.05))$, *and assume* $\max_{s\in K(0.06)} |f(s)| \le 1$. *Then we have*

$$d(\log \zeta, f, K(0.0001), \varepsilon) \ge \exp(-1/\varepsilon^{13}),$$
$$\tau(\log \zeta, f, K(0.0001), \varepsilon) \le \exp\exp(10/\varepsilon^{13}).$$

Besides the above work of Good and Garunkštis, there are various different approaches toward quantitative results. Laurinčikas [81] pointed out that quantitative information on the speed of convergence of a certain functional limit theorem would give a quantitative result on the universality for Lerch zeta-functions. J. Steuding [199] [200] considered the quantity defined by replacing liminf on (1.3) by limsup, and discussed its upper bounds.

The author pointed out in [137] that from Theorem 1.2, by comparing the Taylor expansions of $\zeta(s+i\tau)$ and $f(s)$, it is possible to deduce a certain weaker version of universal approximation. On the other hand, a quantitative version of Theorem 1.2 was shown by Voronin himself [210]. Combining these two ideas, a quantitative version of weak universal approximation theorem was obtained in Garunkštis, Laurinčikas, Matsumoto, J. & R. Steuding [37].

A nice survey on the effectivization problem is given in Laurinčikas [99].

17. The universality for derived functions

When some function $\varphi(s)$ satisfies the universality property, a natural question is to ask whether functions derived from $\varphi(s)$ by some standard operations, such as $\varphi'(s)$, $\varphi(s)^2$, $\exp(\varphi(s))$ etc, also satisfy the universality property, or not.

We already mentioned the universality of $\log\zeta(s)$ (see Remark 16.1). Concerning the derivatives, Bagchi [4] proved that mth derivatives of Dirichlet L-functions $L^{(m)}(s,\chi)$ ($m \in \mathbb{N}$) are strongly universal. Laurinčikas [73] studied the universality for $(\zeta'/\zeta)(s)$, and then considered the same problem for $L(s,g)$ (for a cusp form g) in [86] [87]. The universality for derivatives of L-functions of elliptic curves was studied by Garbaliauskienė and Laurinčikas [32], and its discrete analogue was discussed by Belovas, Garbaliauskienė and Ivanauskaitė [8].

After these early attempts, Laurinčikas [94] (see also [97]) formulated a more general framework of *composite universality*. Let F be an operator $F : H(D(1/2,1)) \to H(D(1/2,1))$. Laurinčikas [94] considered when $F(\zeta(s))$ has the universality property.

A simple affirmative case is the Lipschitz class $\mathrm{Lip}(\alpha)$. We call F belongs to $\mathrm{Lip}(\alpha)$ when

1) for any polynomial $q = q(s)$ and any compact $K \subset D(1/2,1)$, there exists $q_0 \in F^{-1}\{q\}$ such that $q_0(s) \neq 0$ on K, and

2) for any compact $K \subset D(1/2,1)$ with connected complement, there exist $c > 0$ and a compact $K_1 \subset D(1/2,1)$ with connected complement, for which

$$\sup_{s\in K} |F(g_1(s)) - F(g_2(s))| \leq c \sup_{s\in K_1} |g_1(s) - g_2(s)|^{\alpha}$$

holds for all $g_1, g_2 \in H(D(1/2,1))$.

Then it is pointed out in [94] that if $F \in \mathrm{Lip}(\alpha)$, then $F(\zeta(s))$ has the universality property. This claim especially includes the proof of the universality for $\zeta'(s)$.

To prove other theorems in [94], Laurinčikas applied the method of functional limit theorems (cf. Mauclaire [140]). Those results imply, for example, $\zeta'(s) + \zeta''(s)$, $\zeta(s)^m$ (m-th power), $\exp(\zeta(s))$, and $\sin(\zeta(s))$ are universal.

Later, Laurinčikas and his colleagues generalized the result in [94] to various other situations.

• Hurwitz zeta-functions (Laurinčikas [98] for the single case, Laurinčikas [95] for the joint case, and Laurinčikas and Rašytė [120] for the discrete case),

• Periodic and periodic Hurwitz zeta-functions (Laurinčikas [96], Korsakienė, Pocevičienė and Šiaučiūnas [67]),

• Automorphic L-functions (Laurinčikas, Matsumoto and J. Steuding [118]).

A hybrid version of the joint composite universality for Dirichlet L-functions was given by Laurinčikas, Matsumoto and J. Steuding [119].

An alternative approach to composite universality was done by Meyrath [143].

Yet another approach is due to Christ, J. Steuding and Vlachou [20]. Let $\Omega_0 \times \cdots \times \Omega_n$ be an open subset of $\mathbb{C}^{n+1}$, and let $F : \Omega_0 \times \cdots \times \Omega_n \to \mathbb{C}$ be continuous. In [20], the universality of $F(\zeta(s), \zeta'(s), \ldots, \zeta^{(n)}(s))$ was discussed. First, applying the idea in [37], they proved a weaker form of universal approximation. Then, when F is non-constant and analytic, they obtained a kind of universality theorem on a certain small circle. Their proof relies on the implicit function theorem and the Picard-Lindelöf theorem on certain differential equations.

18. Ergodicity and the universality

It is quite natural to understand universality from the ergodic viewpoint. In fact, the universality theorem for a certain Dirichlet series $\varphi(s)$ implies that the orbit $\{\varphi(s + i\tau) \mid \tau \in \mathbb{R}\}$ is dense in a certain function space, and this orbit comes back to an arbitrarily small neighbourhood of any target function infinitely often. This is really an ergodic phenomenon.

Therefore, we can expect that there are some explicit connections between universality theory and ergodic theory. We mentioned already in Section 10 that the Birkhoff-Khinchin theorem in ergodic theory is used in Bagchi's proof of the universality.

Recently J. Steuding [202] formulated a kind of universality theorem, whose statement itself is written in terms of ergodic theory.

Let $(X, \mathcal{B}, P)$ be a probability space, and let $T : X \to X$ a measure-preserving transformation. We call T ergodic with respect to P if $A \in \mathcal{B}$ satisfies $T^{-1}(A) = A$, then either $P(A) = 0$ or $P(A) = 1$ holds. In this case we call $(X, \mathcal{B}, P, T)$ an ergodic dynamical system. Here we consider the case $X = \mathbb{R}$ and $\mathcal{B}$ is the standard Borel σ-algebra.

Let $D \subset \mathbb{C}$ be a domain, $K_1, \ldots, K_r$ be compact subsets of D with connected complements, and $f_j \in H_0^c(K_j)$ $(1 \leq j \leq r)$. We call $\varphi_1, \ldots, \varphi_r \in H(D)$ is *jointly ergodic universal* if for any K_j, f_j, T, $\varepsilon > 0$, and for almost all $x \in \mathbb{R}$, there exists an $n \in \mathbb{N}$ for which

$$\max_{1 \leq j \leq r} \sup_{s \in K_j} |\varphi_j(s + iT^n x) - f_j(s)| < \varepsilon \tag{18.1}$$

holds. Here, $T^n x$ means the T-times iteration of T. If the above statament holds for $f_j \in H^c(K_j)$ $(1 \leq j \leq r)$, then we call $\varphi_1, \ldots, \varphi_r \in H(D)$ *jointly strongly ergodic universal.*

Theorem 18.1. *(J. Steuding [202]) Let D, K_j, T be as above. Let $\varphi_1, \ldots, \varphi_r$ be a family of L-functions. Then, there exists a real number τ such that*

$$\max_{1 \leq j \leq r} \sup_{s \in K_j} |\varphi_j(s + i\tau) - f_j(s)| < \varepsilon \tag{18.2}$$

for any $f_j \in H_0^c(K_j)$ (resp. $H^c(K_j)$) *and any $\varepsilon > 0$, if and only if $\varphi_1, \ldots, \varphi_r$ is jointly (resp. jointly strongly) ergodic universal. And in this case, we have*

$$\liminf_{N \to \infty} \frac{1}{N} \# \left\{ n \in \mathbb{N} \;\middle|\; \max_{1 \leq j \leq r} \sup_{s \in K_j} |\varphi_j(s + iT^n x) - f_j(s)| < \varepsilon \right\} > 0. \tag{18.3}$$

Srichan, R. & J. Steuding [195] discovered the universality produced by a random walk. They considered a lattice Λ on $\mathbb{C}$ and a random walk $(s_n)_{n=0}^{\infty}$ on this lattice, and proved the following result. Let K be a compact subset of $D(1/2, 1)$ with connected complement (with a condition given in terms of Λ), and $f \in H_0^c(K)$. Then for any $\varepsilon > 0$,

$$\liminf_{N \to \infty} \frac{1}{N} \# \left\{ n \in \mathbb{N} \;\middle|\; \sup_{s \in K} |\zeta(s + s_n) - f(s)| < \varepsilon \right\} > 0 \tag{18.4}$$

holds almost surely. They also mentioned a result similar to the above for two-dimensional Brownian motions.

The results in this section suggest that universality is a kind of ergodic phenomenon and is to be understood from the viewpoint of dynamical systems. Universality theorems imply that the properties of zeta-functions in the critical strip are quite inaccessible, which is probably the underlying reason of the extreme difficulty of the Riemann hypothesis. Moreover in Section 9 we mentioned that the Riemann hypothesis itself can be reformulated in terms of dynamical systems. Therefore the Riemann hypothesis is perhaps to be understood as a phenomenon with dynamical-system flavour.[§§] In order to pursue this viewpoint, it is indispensable to study universality more deeply and extensively.

References

1. J. Andersson, Disproof of some conjectures of K. Ramachandra, Hardy-Ramanujan J. **22** (1999), 2-7.
2. R. Ayoub, S. Chowla and H. Walum, On sums involving quadratic characters, J. London Math. Soc. **42** (1967), 152-154.
3. B. Bagchi, The statistical behaviour and universality properties of the Riemann zeta-function and other allied Dirichlet series, Thesis, Calcutta, Indian Statistical Institute, 1981.
4. B. Bagchi, A joint universality theorem for Dirichlet L-functions, Math. Z. **181** (1982), 319-334.
5. B. Bagchi, Recurrence in topological dynamics and the Riemann hypothesis, Acta Math. Hung. **50** (1987), 227-240.
6. R. Balasubramanian and K. Ramachandra, On Riemann zeta-function and allied questions II, Hardy-Ramanujan J. **18** (1995), 10-22.
7. H. Bauer, The value distribution of Artin L-series and zeros of zeta-functions, J. Number Theory **98** (2003), 254-279.
8. I. Belovas, V. Garbaliauskienė and R. Ivanauskaitė, The discrete universality of the derivatives of L-functions of elliptic curves, Šiauliai Math. Semin. **3(11)** (2008), 53-59.
9. G. D. Birkhoff, Démonstration d'un théorème élémentaire sur les fonctions entières, C. R. Acad. Sci. Paris **189** (1929), 473-475.
10. K. M. Bitar, N. N. Khuri and H. C. Ren, Path integrals and Voronin's theorem on the universality of the Riemann zeta function, Ann. Phys. **211** (1991), 172-196.
11. H. Bohr, Zur Theorie der Riemann'schen Zetafunktion im kritischen Streifen, Acta Math. **40** (1915), 67-100.
12. H. Bohr, Über eine quasi-periodische Eigenschaft Dirichletscher Reihen mit Anwendung auf die Dirichletschen L-Funktionen, Math. Ann. **85** (1922), 115-122.
13. H. Bohr and R. Courant, Neue Anwendungen der Theorie der Diophantischen Approximationen auf die Riemannschen Zetafunktion, J. reine Angew. Math. **144** (1914), 249-274.
14. H. Bohr and B. Jessen, Über die Werteverteilung der Riemannschen Zetafunktion, Erste Mitteilung, Acta Math. **54** (1930), 1-35; Zweite Mitteilung, ibid. **58** (1932), 1-55.
15. V. Borchsenius and B. Jessen, Mean motions and values of the Riemann zeta function, Acta Math. **80** (1948), 97-166.

[§§]This argument might remind us the work of Deninger [21] [22] which is in a different context but also with dynamical-system flavour.

16. E. Buivydas and A. Laurinčikas, A discrete version of the Mishou theorem, preprint.
17. E. Buivydas and A. Laurinčikas, A discrete version of the joint universality theorem for the Riemann and Hurwitz zeta-functions, preprint.
18. E. Buivydas, A. Laurinčikas, R. Macaitienė and J. Rašytė, Discrete universality theorems for the Hurwitz zeta-function, J. Approx. Theory **183** (2014), 1-13.
19. Y. Cai, Prime geodesic theorem, J. Théor. Nombr. Bordeaux **14** (2002), 59-72.
20. T. Christ, J. Steuding and V. Vlachou, Differential universality, Math. Nachr. **286** (2013), 160-170.
21. C. Deninger, Some analogies between number theory and dynamical systems on foliated spaces, in "Proc. Intern. Congr. Math. Berlin 1998", Vol. I, Documenta Math. J. DMV Extra Vol., 1998, pp.163-186.
22. C. Deninger, On dynamical systems and their possible significance for arithmetic geometry, in "Regulators in Analysis, Geometry and Number Theory", A. Reznikov and N. Schappacher (eds.), Progr. in Math. **171**, Birkhäuser, 2000, pp.29-87.
23. P. Drungilas, R. Garunkštis and A. Kačėnas, Universality of the Selberg zeta-function for the modular group, Forum Math. **25** (2013), 533-564.
24. A. Dubickas, On the linear independence of the set of Dirichlet exponents, Kodai Math. J. **35** (2012), 642-651.
25. K. M. Eminyan, χ-universality of the Dirichlet L-function, Mat. Zametki **47** (1990), 132-137 (in Russian); Math. Notes **47** (1990), 618-622.
26. V. Garbaliauskienė, A weighted universality theorem for zeta-functions of elliptic curves, Liet. Mat. Rink. **44**, Spec. Issue (2004), 43-47.
27. V. Garbaliauskienė, A weighted discrete universality theorem for L-functions of elliptic curves, Liet. Mat. Rink. **45**, Spec. Issue (2005), 25-29.
28. V. Garbaliauskienė, J. Genys and A. Laurinčikas, Discrete universality of the L-functions of elliptic curves, Sibirskiĭ Mat. Zh. **49** (2008), 768-785 (in Russian); Siberian Math. J. **49** (2008), 612-627.
29. V. Garbaliauskienė, R. Kačinskaitė and A. Laurinčikas, The joint universality for L-functions of elliptic curves, Nonlinear Anal. Modell. Control **9** (2004), 331-348.
30. V. Garbaliauskienė and A. Laurinčikas, Discrete value-distribution of L-functions of elliptic curves, Publ. Inst. Math. (Beograd) **76(90)** (2004), 65-71.
31. V. Garbaliauskienė and A. Laurinčikas, Some analytic properties for L-functions of elliptic curves, Proc. Inst. Math. NAN Belarus **13** (2005), 75-82.
32. V. Garbaliauskienė and A. Laurinčikas, The universality of the derivatives of L-functions of elliptic curves, in "Analytic and Probabilistic Methods in Number Theory" (Proc. 4th Palanga Conf.), A. Laurinčikas and E. Manstavičius (eds.), TEV, 2007, pp.24-29.
33. R. Garunkštis, The effective universality theorem for the Riemann zeta function, in "Proc. Session in Analytic Number Theory and Diophantine

Equations", D. R. Heath-Brown and B. Z. Moroz (eds.), Bonner Math. Schriften **360**, Bonn, 2003, n.16, 21pp.
34. R. Garunkštis, Self-approximation of Dirichlet L-functions, J. Number Theory **131** (2011), 1286-1295.
35. R. Garunkštis and E. Karikovas, Self-approximation of Hurwitz zeta-functions, Funct. Approx. Comment. Math. **51** (2014), 181–188.
36. R. Garunkštis, A. Laurinčikas, R. Šleževičienė and J. Steuding, On the universality of Estermann zeta-functions, Analysis **22** (2002), 285-296.
37. R. Garunkštis, A. Laurinčikas, K. Matsumoto, J. Steuding and R. Steuding, Effective uniform approximation by the Riemann zeta-function, Publ. Math. (Barcelona) **54** (2010), 209-219.
38. J. Genys and A. Laurinčikas, Value distribution of general Dirichlet series V, Liet. Mat. Rink. **44** (2004), 181-195 (in Russian); Lith. Math. J. **44** (2004), 145-156.
39. J. Genys, R. Macaitienė, S. Račkauskienė and D. Šiaučiūnas, A mixed joint universality theorem for zeta-functions, Math. Modell. Anal. **15** (2010), 431-446.
40. S. M. Gonek, Analytic properties of zeta and L-functions, Thesis, University of Michigan, 1979.
41. A. Good, On the distribution of the values of Riemann's zeta-function, Acta Arith. **38** (1981), 347-388.
42. K.-G. Grosse-Erdmann, Holomorphe Monster und universelle Funktionen, Mitt. Math. Sem. Giessen **176** (1987), 1-81.
43. K.-G. Grosse-Erdmann, Universal families and hypercyclic operators, Bull. Amer. Math. Soc. **36** (1999), 345-381.
44. M. C. Gutzwiller, Stochastic behavior in quantum scattering, Physica **7D** (1983), 341-355.
45. D. Hilbert, Mathematische Probleme, Nachr. Königl. Ges. Wiss. Göttingen, Math.-phys. Kl. (1900), 253-297; reprinted in Arch. Math. Phys. **(3)1** (1901), 44-63, 213-237; also in Hilbert's Gesammelte Abhandlungen, Vol. III, Chelsea, 1965 (originally in 1935), pp.290-329.
46. J. Ignatavičiūtė, Discrete universality of the Lerch zeta-function, in "Abstracts, 8th Vilnius Conf. on Probab. Theory", TEV, 2002, pp.116-117.
47. K. Janulis and A. Laurinčikas, Joint universality of Dirichlet L-functions and Hurwitz zeta-functions, Ann. Univ. Sci. Budapest., Sect. Comp. **39** (2013), 203-214.
48. K. Janulis, A. Laurinčikas, R. Macaitienė and D. Šiaučiūnas, Joint universality of Dirichlet L-functions and periodic Hurwitz zeta-functions, Math. Modell. Anal. **17** (2012), 673-685.
49. A. Javtokas and A. Laurinčikas, On the periodic Hurwitz zeta-function, Hardy-Ramanujan J. **29** (2006), 18-36.
50. A. Javtokas and A. Laurinčikas, Universality of the periodic Hurwitz zeta-function, Integr. Transf. Spec. Funct. **17** (2006), 711-722.
51. A. Javtokas and A. Laurinčikas, A joint universality theorem for periodic Hurwitz zeta-functions, Bull. Austral. Math. Soc. **78** (2008), 13-33.

52. B. Jessen and H. Tornehave, Mean motions and zeros of almost periodic functions, Acta Math. **77** (1945), 137-279.
53. A. Kačėnas and A. Laurinčikas, On Dirichlet series related to certain cusp forms, Liet. Mat. Rink. **38** (1998), 82-97 (in Russian); Lith. Math. J. **38** (1998), 64-76.
54. R. Kačinskaitė, A discrete universality theorem for the Matsumoto zeta-function, Liet. Mat. Rink. **42**, Spec. Issue (2002), 55-58.
55. R. Kačinskaitė, Joint discrete universality of periodic zeta-functions, Integr. Transf. Spec. Funct. **22** (2011), 593-601.
56. R. Kačinskaitė, Limit theorems for zeta-functions — with application in universality, Šiauliai Math. Semin. **7(15)** (2012), 19-40.
57. R. Kačinskaitė, A. Javtokas and D. Šiaučiūnas, On discrete universality of the periodic zeta-function, Šiauliai Math. Semin. **3(11)** (2008), 141-152.
58. R. Kačinskaitė and A. Laurinčikas, The joint distribution of periodic zeta-functions, Studia Sci. Math. Hung. **48** (2011), 257-279.
59. R. Kačinskaitė, J. Steuding, D. Šiaučiūnas and R. Šleževičienė, On polynomials in Dirichlet series, Fiz. Mat. Fak. Moksl. Sem. Darbai, Šiauliai Univ. **7** (2004), 26-32.
60. J. Kaczorowski, Some remarks on the universality of periodic L-functions, in "New Directions in Value-Distribution Theory of Zeta and L-Functions", R.& J. Steuding (eds.), Shaker Verlag, 2009, pp.113-120.
61. J. Kaczorowski and M. Kulas, On the non-trivial zeros off the critical line for L-functions from the extended Selberg class, Monatsh. Math. **150** (2007), 217-232.
62. J. Kaczorowski, A. Laurinčikas and J. Steuding, On the value distribution of shifts of universal Dirichlet series, Monatsh. Math. **147** (2006), 309-317.
63. J. Kaczorowski and A. Perelli, The Selberg class: a survey, in "Number Theory in Progress", Proc. Intern. Conf. on Number Theory in Honor of the 60th Birthday of A. Schinzel at Zakopane, Vol. 2, Elementary and Analytic Number Theory, K. Györy et al. (eds.), Walter de Gruyter, 1999, pp.953-992.
64. A. A. Karatsuba and S. M. Voronin, The Riemann Zeta-Function, Walter de Gruyter, 1992.
65. E. Karikovas and Ł. Pańkowski, Self-approximation of Hurwitz zeta-functions with rational parameter, Lith. Math. J. **54** (2014), 74-81.
66. H. Ki and Y. Lee, On the zeros of degree one L-functions from the extended Selberg class, Acta Arith. **149** (2011), 23-36.
67. D. Korsakienė, V. Pocevičienė and D. Šiaučiūnas, On universality of periodic zeta-functions, Šiauliai Math. Semin. **8(16)** (2013), 131-141.
68. A. Laurinčikas, Distribution des valeurs de certaines séries de Dirichlet, C. R. Acad. Sci. Paris **289** (1979), 43-45.
69. A. Laurinčikas, Sur les séries de Dirichlet et les polynômes trigonométriques, Sém. Théor. Nombr., Univ. de Bordeaux I, Éxposé no.24, 1979.
70. A. Laurinčikas, Distribution of values of generating Dirichlet series of multiplicative functions, Liet. Mat. Rink. **22** (1982), 101-111 (in Russian); Lith. Math. J. **22** (1982), 56-63.

71. A. Laurinčikas, The universality theorem, Liet. Mat. Rink. **23** (1983), 53-62 (in Russian); Lith. Math. J. **23** (1983), 283-289.
72. A. Laurinčikas, The universality theorem II, Liet. Mat. Rink. **24** (1984), 113-121 (in Russian); Lith. Math. J. **24** (1984), 143-149.
73. A. Laurinčikas, Zeros of the derivative of the Riemann zeta-function, Liet. Mat. Rink. **25** (1985), 111-118 (in Russian); Lith. Math. J. **25** (1985), 255-260.
74. A. Laurinčikas, Zeros of linear combinations of Dirichlet series, Liet. Mat. Rink. **26** (1986), 468-477 (in Russian); Lith. Math. J. **26** (1986), 244-251.
75. A. Laurinčikas, On the universality of the Riemann zeta-function, Liet. Mat. Rink. **35** (1995), 502-507 (in Russian); Lith. Math. J. **35** (1995), 399-402.
76. A. Laurinčikas, Limit Theorems for the Riemann Zeta-function, Kluwer, 1996.
77. A. Laurinčikas, The universality of the Lerch zeta-function, Liet. Mat. Rink. **37** (1997), 367-375 (in Russian); Lith. Math. J. **37** (1997), 275-280.
78. A. Laurinčikas, On the Matsumoto zeta-function, Acta Arith. **84** (1998), 1-16.
79. A. Laurinčikas, On the zeros of linear combinations of the Matsumoto zeta-functions, Liet. Mat. Rink. **38** (1998), 185-204 (in Russian); Lith. Math. J. **38** (1998), 144-159.
80. A. Laurinčikas, On the Lerch zeta-function with rational parameters, Liet. Mat. Rink. **38** (1998), 113-124 (in Russian); Lith. Math. J. **38** (1998), 89-97.
81. A. Laurinčikas, On the effectivization of the universality theorem for the Lerch zeta-function, Liet. Mat. Rink. **40** (2000), 172-178 (in Russian); Lith. Math. J. **40** (2000), 135-139.
82. A. Laurinčikas, The universality of Dirichlet series attached to finite Abelian groups, in "Number Theory", M. Jutila and T. Metsänkylä (eds.), Walter de Gruyter, 2001, 179-192.
83. A. Laurinčikas, The universality of zeta-functions, Acta Appl. Math. **78** (2003), 251-271.
84. A. Laurinčikas, The joint universality for general Dirichlet series, Ann. Univ. Sci. Budapest., Sect. Comp. **22** (2003), 235-251.
85. A. Laurinčikas, Joint universality of general Dirichlet series, Izv. Ross. Akad. Nauk Ser. Mat. **69** (2005), 133-144 (in Russian); Izv. Math. **69** (2005), 131-142.
86. A. Laurinčikas, On the derivatives of zeta-functions of certain cusp forms, Glasgow Math. J. **47** (2005), 87-96.
87. A. Laurinčikas, On the derivatives of zeta-functions of certain cusp forms II, Glasgow Math. J. **47** (2005), 505-516.
88. A. Laurinčikas, The joint universality for periodic Hurwitz zeta-functions, Analysis **26** (2006), 419-428.
89. A. Laurinčikas, Voronin-type theorem for periodic Hurwitz zeta-functions, Mat. Sb. **198** (2007), 91-102 (in Russian); Sb. Math. **198** (2007), 231-242.

90. A. Laurinčikas, Joint universality for periodic Hurwitz zeta-functions, Izv. Ross. Akad. Nauk Ser. Mat. **72** (2008), 121-140 (in Russian); Izv. Math. **72** (2008), 741-760.
91. A. Laurinčikas, The joint universality of Hurwitz zeta-functions, Šiauliai Math. Semin. **3(11)** (2008), 169-187.
92. A. Laurinčikas, On the joint universality of Lerch zeta functions, Mat. Zametki **88** (2010), 428-437 (in Russian); Math. Notes **88** (2010), 386-394.
93. A. Laurinčikas, Joint universality of zeta-functions with periodic coefficients, Izv. Ross. Akad. Nauk Ser. Mat. **74** (2010), 79-102 (in Russian); Izv. Math. **74** (2010), 515-539.
94. A. Laurinčikas, Universality of the Riemann zeta-function, J. Number Theory **130** (2010), 2323-2331.
95. A. Laurinčikas, Joint universality of Hurwitz zeta-functions, Bull. Austral. Math. Soc. **86** (2012), 232-243.
96. A. Laurinčikas, Universality of composite functions of periodic zeta functions, Mat. Sb. **203** (2012), 105-120 (in Rissian); Sb. Math. **203** (2012), 1631-1646.
97. A. Laurinčikas, Universality of composite functions, in "Functions in Number Theory and Their Probabilistic Aspects", K. Matsumoto et al. (eds.), RIMS Kôkyûroku Bessatsu **B34**, RIMS, 2012, pp.191-204.
98. A. Laurinčikas, On the universality of the Hurwitz zeta-function, Intern. J. Number Theory **9** (2013), 155-165.
99. A. Laurinčikas, Universality results for the Riemann zeta-function, Moscow J. Combin. Number Theory **3** (2013), 237-256.
100. A. Laurinčikas, A discrete universality theorem for the Hurwitz zeta-function, J. Number Theory, to appear.
101. A. Laurinčikas, Joint discrete universality of Hurwitz zeta-functions, preprint (in Russian).
102. A. Laurinčikas and R. Garunkštis, The Lerch Zeta-function, Kluwer, 2002.
103. A. Laurinčikas and R. Macaitienė, On the joint universality of periodic zeta functions, Mat. Zametki **85** (2009), 54-64 (in Russian); Math. Notes **85** (2009), 51-60.
104. A. Laurinčikas and R. Macaitienė, The discrete universality of the periodic Hurwitz zeta function, Integr. Transf. Spec. Funct. **20** (2009), 673-686.
105. A. Laurinčikas and R. Macaitienė, On the universality of zeta-functions of certain cusp forms, in "Analytic and Probabilistic Methods in Number Theory" (Kubilius Memorial Volume), A. Laurinčikas et al. (eds.), TEV, 2012, pp.173-183.
106. A. Laurinčikas and R. Macaitienė, Joint universality of the Riemann zeta-function and Lerch zeta-functions, Nonlinear Anal. Modell. Control **18** (2013), 314-326.
107. A. Laurinčikas, R. Macaitienė and D. Šiaučiūnas, The joint universality for periodic zeta-functions, Chebyshevskiĭ Sb. **8** (2007), 162-174.
108. A. Laurinčikas, R. Macaitienė and D. Šiaučiūnas, On discrete universality of the periodic zeta-function II, in "New Directions in Value-Distribution

Theory of Zeta and L-Functions", R.& J. Steuding (eds.), Shaker Verlag, 2009, pp.149-159.

109. A. Laurinčikas, R. Macaitienė and D. Šiaučiūnas, Joint universality for zeta-functions of different types, Chebyshevskiĭ Sb. **12** (2011), 192-203.
110. A. Laurinčikas and K. Matsumoto, The joint universality and the functional independence for Lerch zeta-functions, Nagoya Math. J. **157** (2000), 211-227.
111. A. Laurinčikas and K. Matsumoto, The universality of zeta-functions attached to certain cusp forms, Acta Arith. **98** (2001), 345-359.
112. A. Laurinčikas and K. Matsumoto, The joint universality of zeta-functions attached to certain cusp forms, Fiz. Mat. Fak. Moksl. Sem. Darbai, Šiauliai Univ. **5** (2002), 58-75.
113. A. Laurinčikas and K. Matsumoto, The joint universality of twisted automorphic L-functions, J. Math. Soc. Japan **56** (2004), 923-939.
114. A. Laurinčikas and K. Matsumoto, Joint value-distribution theorems on Lerch zeta-functions II, Liet. Mat. Rink. **46** (2006), 332-350; Lith. Math. J. **46** (2006), 271-286.
115. A. Laurinčikas and K. Matsumoto, Joint value-distribution theorems on Lerch zeta-functions III, in "Analytic and Probabilistic Methods in Number Theory" (Proc. 4th Palanga Conf.), A. Laurinčikas and E. Manstavičius (eds.), TEV, 2007, pp.87-98.
116. A. Laurinčikas, K. Matsumoto and J. Steuding, The universality of L-functions associated with new forms, Izv. Ross. Akad. Nauk Ser. Mat. **67** (2003), 83-98 (in Russian); Izv. Math. **67** (2003), 77-90.
117. A. Laurinčikas, K. Matsumoto and J. Steuding, Discrete universality of L-functions for new forms, Mat. Zametki **78** (2005), 595-603 (in Russian); Math. Notes **78** (2005), 551-558.
118. A. Laurinčikas, K. Matsumoto and J. Steuding, Universality of some functions related to zeta-functions of certain cusp forms, Osaka J. Math. **50** (2013), 1021-1037.
119. A. Laurinčikas, K. Matsumoto and J. Steuding, On hybrid universality of certain composite functions involving Dirichlet L-functions, Ann. Univ. Sci. Budapest., Sect. Comp. **41** (2013), 85-96.
120. A. Laurinčikas and J. Rašytė, Generalizations of a discrete universality theorem for Hurwitz zeta-functions, Lith. Math. J. **52** (2012), 172-180.
121. A. Laurinčikas, W. Schwarz and J. Steuding, The universality of general Dirichlet series, Analysis **23** (2003), 13-26.
122. A. Laurinčikas and D. Šiaučiūnas, Remarks on the universality of the periodic zeta function, Mat. Zametki **80** (2006), 561-568 (in Russian); Math. Notes **80** (2006), 532-538.
123. A. Laurinčikas and D. Šiaučiūnas, A mixed joint universality theorem for zeta-functions III, in "Analytic and Probabilistic Methods in Number Theory" (Kubilius Memorial Volume), A. Laurinčikas et al. (eds.), TEV, 2012, pp.185-195.

124. A. Laurinčikas and S. Skerstonaitė, A joint universality theorem for periodic Hurwitz zeta-functions II,¶¶ Lith. Math. J. **49** (2009), 287-296.
125. A. Laurinčikas and S. Skerstonaitė, Joint universality for periodic Hurwitz zeta-functions II, in "New Directions in Value-Distribution Theory of Zeta and L-Functions", R.& J. Steuding (eds.), Shaker Verlag, 2009, pp.161-169.
126. A. Laurinčikas and R. Šleževičienė, The universality of zeta-functions with multiplicative coefficients, Integr. Transf. Spec. Funct. **13** (2002), 243-257.
127. Y. Lee, The universality theorem for Hecke L-functions, Math. Z. **271** (2012), 893-909.
128. Y. Lee, Zeros of partial zeta functions off the critical line, in "Functions in Number Theory and Their Probabilistic Aspects", K. Matsumoto et al. (eds.), RIMS Kôkyûroku Bessatsu **B34**, RIMS, 2012, pp.205-216.
129. H. Li and J. Wu, The universality of symmetric power L-functions and their Rankin-Selberg L-functions, J. Math. Soc. Japan **59** (2007), 371-392.
130. R. Macaitienė, A discrete universality theorem for general Dirichlet series, Analysis **26** (2006), 373-381.
131. R. Macaitienė, On joint universality for the zeta-functions of newforms and periodic Hurwitz zeta-functions, in "Functions in Number Theory and Their Probabilistic Aspects", K. Matsumoto et al. (eds.), RIMS Kôkyûroku Bessatsu **B34**, RIMS, 2012, pp.217-233.
132. J. Marcinkiewicz, Sur les nombres dérivés, Fund. Math. **24** (1935), 305-308.
133. K. Matsumoto, Value-distribution of zeta-functions, in "Analytic Number Theory", Proc. Japanese-French Sympos., K. Nagasaka and E. Fouvry (eds.), Lecture Notes in Math. **1434**, Springer, 1990, pp.178-187.
134. K. Matsumoto, The mean values and the universality of Rankin-Selberg L-functions, in "Number Theory", M. Jutila and T. Metsänkylä (eds.), Walter de Gruyter, 2001, 201-221.
135. K. Matsumoto, Some problems on mean values and the universality of zeta and multiple zeta-functions, in "Analytic and Probabilistic Methods in Number Theory" (Proc. 3rd Palanga Conf.), A. Dubickas et al (eds.), TEV, 2002, pp.195-199.
136. K. Matsumoto, Probabilistic value-distribution theory of zeta-functions, Sūgaku **53** (2001), 279-296 (in Japanese); Sugaku Expositions **17** (2004), 51-71.
137. K. Matsumoto, An introduction to the value-distribution theory of zeta-functions, Šiauliai Math. Semin. **1(9)** (2006), 61-83.
138. J.-L. Mauclaire, Almost periodicity and Dirichlet series, in "Analytic and Probabilistic Methods in Number Theory" (Proc. 4th Palanga Conf.), A. Laurinčikas and E. Manstavičius (eds.), TEV, 2007, pp.109-142.
139. J.-L. Mauclaire, On some Dirichlet series, in "New Directions in Value-Distribution Theory of Zeta and L-Functions", R.& J. Steuding (eds.), Shaker Verlag, 2009, pp.171-248.
140. J.-L. Mauclaire, Simple remarks on some Dirichlet series, Ann. Univ. Sci. Budapest., Sect. Comp. **41** (2013), 159-172.

¶¶This "II" is probably to be deleted.

141. S. N. Mergelyan, On the representation of functions by series of polynomials on closed sets, Dokl. Akad. Nauk SSSR **78** (1951), 405-408 (in Russian); Amer. Math. Soc. Transl. Ser.1, **3**, Series and Approximation, Amer. Math. Soc., 1962, pp.287-293.
142. S. N. Mergelyan, Uniform approximation to functions of complex variable, Usp. Mat. Nauk **7** (1952), 31-122 (in Russian); Amer. Math. Soc. Transl. Ser.1, **3**, Series and Approximation, Amer. Math. Soc., 1962, pp.294-391.
143. T. Meyrath, On the universality of derived functions of the Riemann zeta-function, J. Approx. Theory **163** (2011), 1419-1426.
144. H. Mishou, The universality theorem for L-functions associated with ideal class characters, Acta Arith. **98** (2001), 395-410.
145. H. Mishou, The universality theorem for Hecke L-functions, Acta Arith. **110** (2003), 45-71.
146. H. Mishou, On the value distribution of Hecke L-functions associated with grössencharacters, Sûrikaiseki Kenkyûsho Kôkyûroku **1324** (2003), 174-182 (in Japanese).
147. H. Mishou, The value distribution of Hecke L-functions in the grössencharacter aspect, Arch. Math. **82** (2004), 301-310.
148. H. Mishou, The universality theorem for Hecke L-functions in the (m, t) aspect, Tokyo J. Math. **28** (2005), 139-153.
149. H. Mishou, The joint value-distribution of the Riemann zeta function and Hurwitz zeta functions, Liet. Mat. Rink. **47** (2007), 62-80; Lith. Math. J. **47** (2007), 32-47.
150. H. Mishou, The universality theorem for cubic L-functions, in "New Directions in Value-Distribution Theory of Zeta and L-Functions", R.& J. Steuding (eds.), Shaker Verlag, 2009, pp.265-274.
151. H. Mishou, The universality theorem for class group L-functions, Acta Arith. **147** (2011), 115-128.
152. H. Mishou, The joint universality theorem for a pair of Hurwitz zeta functions, J. Number Theory **131** (2011), 2352-2367.
153. H. Mishou, On joint universality for derivatives of the Riemann zeta function and automorphic L-functions, in "Functions in Number Theory and Their Probabilistic Aspects", K. Matsumoto et al. (eds.), RIMS Kôkyûroku Bessatsu **B34**, RIMS, 2012, pp.235-246.
154. H. Mishou, Joint value distribution for zeta functions in disjoint strips, Monatsh. Math. **169** (2013), 219-247.
155. H. Mishou, Functional distribution for a collection of Lerch zeta functions, J. Math. Soc. Japan **66** (2014), 1105-1126.
156. H. Mishou, Joint universality theorems for pairs of automorphic zeta functions, Math. Z. **277** (2014), 1113-1154.
157. H. Mishou and S. Koyama, Universality of Hecke L-functions in the Grossencharacter-aspect, Proc. Japan Acad. **78A** (2002), 63-67.
158. H. Mishou and H. Nagoshi, Functional distribution of $L(s, \chi_d)$ with real characters and denseness of quadratic class numbers, Trans. Amer. Math. Soc. **358** (2006), 4343-4366.

159. H. Mishou and H. Nagoshi, The universality of quadratic L-series for prime discriminants, Acta Arith. **123** (2006), 143-161.
160. H. Mishou and H. Nagoshi, Equivalents of the Riemann hypothesis, Arch. Math. **86** (2006), 419-424.
161. H. Mishou and H. Nagoshi, On class numbers of quadratic fields with prime discriminant and character sums, Kyushu J. Math. **66** (2012), 21-34.
162. H. L. Montgomery, Extreme values of the Riemann zeta-function, Comment. Math. Helv. **52** (1977), 511-518.
163. H. Nagoshi, On the universality for L-functions attached to Maass forms, Analysis **25** (2005), 1-22.
164. H. Nagoshi, The universality of L-functions attached to Maass forms, in "Probability and Number Theory — Kanazawa 2005", S. Akiyama et al. (eds.), Adv. Stud. Pure Math. **49**, Math. Soc. Japan, 2007, pp.289-306.
165. H. Nagoshi, Value-distribution of Rankin-Selberg L-functions, in "New Directions in Value-Distribution Theory of Zeta and L-Functions", R.& J. Steuding (eds.), Shaker Verlag, 2009, pp.275-287.
166. H. Nagoshi and J. Steuding, Universality for L-functions in the Selberg class, Lith. Math. J. **50** (2010), 293-311.
167. T. Nakamura, Applications of inversion formulas to the joint t-universality of Lerch zeta functions, J. Number Theory **123** (2007), 1-9.
168. T. Nakamura, The existence and the non-existence of joint t-universality for Lerch zeta functions, J. Number Theory **125** (2007), 424-441.
169. T. Nakamura, Joint value approximation and joint universality for several types of zeta functions, Acta Arith. **134** (2008), 67-82.
170. T. Nakamura, Zeros and the universality for the Euler-Zagier-Hurwitz type of multiple zeta-functions, Bull. London Math. Soc. **41** (2009), 691-700.
171. T. Nakamura, The joint universality and the generalized strong recurrence for Dirichlet L-functions, Acta Arith. **138** (2009), 357-262.
172. T. Nakamura, The generalized strong recurrence for non-zero rational parameters, Arch. Math. **95** (2010), 549-555.
173. T. Nakamura, The universality for linear combinations of Lerch zeta functions and the Tornheim-Hurwitz type of double zeta functions, Monatsh. Math. **162** (2011), 167-178.
174. T. Nakamura, Some topics related to universality of L-functions with an Euler product, Analysis **31** (2011), 31-41.
175. T. Nakamura and Ł. Pańkowski, Applications of hybrid universality to multivariable zeta-functions, J. Number Theory **131** (2011), 2151-2161.
176. T. Nakamura and Ł. Pańkowski, Erratum to "The generalized strong recurrence for non-zero rational parameters", Arch. Math. **99** (2012), 43-47.
177. T. Nakamura and Ł. Pańkowski, On universality for linear combinations of L-functions, Monatsh. Math. **165** (2012), 433-446.
178. T. Nakamura and Ł. Pańkowski, Self-approximation for the Riemann zeta function, Bull. Austral. Math. Soc. **87** (2013), 452-461.
179. T. Nakamura and Ł. Pańkowski, On zeros and c-values of Epstein zeta-functions, Šiauliai Math. Semin. **8(16)** (2013), 181-195.

180. T. Nakamura and Ł. Pańkowski, On complex zeros off the critical line for non-monomial polynomial of zeta-functions, preprint.
181. J. Pál, Zwei kleine Bemerkungen, Tôhoku Math. J. **6** (1914/15), 42-43.
182. Ł. Pańkowski, Some remarks on the generalized strong recurrence for L-functions, in "New Directions in Value-Distribution Theory of Zeta and L-Functions", R.& J. Steuding (eds.), Shaker Verlag, 2009, pp.305-315.
183. Ł. Pańkowski, Hybrid universality theorem for Dirichlet L-functions, Acta Arith. **141** (2010), 59-72.
184. Ł. Pańkowski, Hybrid universality theorem for L-functions without Euler product, Integr. Transf. Spec. Funct. **24** (2013), 39-49.
185. D. V. Pecherskiĭ, On rearrangements of terms in functional series, Dokl. Akad. Nauk SSSR **209** (1973), 1285-1287 (in Russian); Soviet Math. Dokl. **14** (1973), 633-636.
186. V. Pocevičienė and D. Šiaučiūnas, A mixed joint universality theorem for zeta-functions II, Math. Modell. Anal. **19** (2014), 52-65.
187. K. Ramachandra, On Riemann zeta-function and allied questions, in "Journées Arithmétiques de Genève", D. F. Coray and Y.-F. S. Pétermann (eds.), Astérisque **209**, Soc. Math. France, 1992, pp.57-72.
188. A. Reich, Universalle Werteverteilung von Eulerprodukten, Nachr. Akad. Wiss. Göttingen II Math.-Phys. Kl., Nr.1 (1977), 1-17.
189. A. Reich, Werteverteilung von Zetafunktionen, Arch. Math. **34** (1980), 440-451.
190. A. Reich, Zur Universität und Hypertranszendenz der Dedekindschen Zetafunktion, Abh. Braunschweig. Wiss. Ges. **33** (1982), 197-203.
191. J. Sander and J. Steuding, Joint universality for sums and products of Dirichlet L-functions, Analysis **26** (2006), 295-312.
192. W. Schwarz, R. Steuding and J. Steuding, Universality for Euler products, and related arithmetical functions, in "Analytic and Probabilistic Methods in Number Theory" (Proc. 4th Palanga Conf.), A. Laurinčikas and E. Manstavičius (eds.), TEV, 2007, pp.163-189.
193. A. Selberg, Old and new conjectures and results about a class of Dirichlet series, in "Proceedings of the Amalfi Conference on Analytic Number Theory", E. Bombieri et al. (eds.), Univ. di Salerno, 1992, pp.367-385; also in Selberg's Collected Papers, Vol. II, Springer, 1991, pp.47-63.
194. R. Šleževičienė,*** The joint universality for twists of Dirichlet series with multiplicative coefficients by characters, in "Analytic and Probabilistic Methods in Number Theory" (Proc. 3rd Palanga Conf.), A. Dubickas et al (eds.), TEV, 2002, pp.303-319.
195. T. Srichan, R. Steuding and J. Steuding, Does a random walker meet universality? Šiauliai Math. Semin. **8(16)** (2013), 249-259.
196. J. Steuding, The world of p-adic numbers and p-adic functions, Fiz. Mat. Fak. Moksl. Sem. Darbai, Šiauliai Univ. **5** (2002), 90-107.
197. J. Steuding, On the universality for functions in the Selberg class, in "Proc. Session in Analytic Number Theory and Diophantine Equations", D. R.

***R. Šleževičienė= R. Steuding

Heath-Brown and B. Z. Moroz (eds.), Bonner Math. Schriften **360**, Bonn, 2003, n.28, 22pp.
198. J. Steuding, Value-distribution of L-functions and allied zeta-functions — with an emphasis on aspects of universality, Habilitationsschrift, Frankfurt, Johann Wolfgang Goethe Universität, 2003.
199. J. Steuding, Upper bounds for the density of universality, Acta Arith. **107** (2003), 195-202.
200. J. Steuding, Upper bounds for the density of universality II, Acta Math. Univ. Ostrav. **13** (2005), 73-82.
201. J. Steuding, Value-distribution of L-functions, Lecture Notes in Math. **1877**, Springer, 2007.
202. J. Steuding, Ergodic universality theorems for Riemann's zeta-function and other L-functions, J. Théor. Nombr. Bordeaux **25** (2013), 471-476.
203. R. Steuding, Universality for generalized Euler products, Analysis **26** (2006), 337-345.
204. S. M. Voronin, On the distribution of nonzero values of the Riemann ζ-function, Trudy Mat. Inst. Steklov. **128** (1972), 131-150 (in Russian); Proc. Steklov Inst. Math. **128** (1972), 153-175.
205. S. M. Voronin, Theorem on the "universality' of the Riemann zeta-function, Izv. Akad. Nauk SSSR Ser. Mat. **39** (1975), 475-486 (in Russian); Math. USSR Izv. **9** (1975), 443-453.
206. S. M. Voronin, On the functional independence of Dirichlet L-functions, Acta Arith. **27** (1975), 443-453 (in Russian).
207. S. M. Voronin, Analytic properties of Dirichlet generating functions of arithmetic objects, Thesis, Moscow, Steklov Math. Institute, 1977 (in Russian).
208. S. M. Voronin, Analytic properties of Dirichlet generating functions of arithmetic objects, Mat. Zametki **24** (1978), 879-884 (in Russian); Math. Notes **24** (1979), 966-969.
209. S. M. Voronin, On the distribution of zeros of some Dirichlet series, Trudy Mat. Inst. Steklov. **163** (1984), 74-77 (in Russian); Proc. Steklov Inst. Math. (1985), Issue 4, 89-92.
210. S. M. Voronin, On Ω-theorems in the theory of the Riemann zeta-function, Izv. Akad. Nauk SSSR Ser. Mat. **52** (1988), 424-436 (in Russian); Math. USSR Izv. **32** (1989), 429-442.

COMPLEX MULTIPLICATION IN THE SENSE OF ABEL

KATSUYA MIYAKE

Emeritus Professor of Tokyo Metropolitan University, Japan
Visit. Prof. of Inst. for Math. and Comp. Sci., Tsuda College, Japan
E-mail: miyakek@bz-csp.tepm.jp

1. The origins of complex multiplication

1.1. *Introduction*

Complex multiplication in the modern sense is a certain property of an Abelian variety. In this paper we visit the origin in the celebrated paper of Abel on elliptic functions [Ab-1827],

Recherches sur les fonctions elliptiques (Research on elliptic functions), which was published in 1827.

The words 'complex multiplication' actually first appeared in 1857 in the title of Kronecker's paper [Kr-1857a],

Über die elliptische Funktionen, für welche complex Multiplication stattfindet

(On elliptic functions for which complex multiplication occurs).

Here we should mention the different meanings of the concepts of 'elliptic functions' in the titles of these two papers. As was mentioned in [Mi-2011, p.135] for example, Abel used the term 'fonctions elliptiques' for the integral functions of elliptic integrals after Legendre.

Legendre systematically investigated elliptic integrals ([Le-1793, -1811, -1825]), reduced them to three types, the first, the second and the third, and published tables of the values of them ([Le-1811]). Legendre never dared to step out of the world of real numbers by himself. His works were, however, steady stepping stones for the young researchers. Abel started, also strongly influenced by the inspiring work of Gauss [Ga-1801], *Disquisitiones arithmeticae*, to investigate elliptic functions (in the sense of old Legendre) in the wide world of complex numbers and their inverse functions of one complex variable; he found, for example, double periodicity of the latter which could only be realized on the complex plane.

It was Carl Gustav Jacobi who introduced the term 'elliptic integrals' and called the inverse functions of the complex elliptic integral functions 'elliptic functions' in 1829 for the memory of Abel who passed away too early; see [Ja-1829a, Ja-1829b]. He dared propose the change of the meanings of the term 'elliptic functions' which had been used since Legendre's work [Le-1811] of 1811, and won the approval of the big old mathematician. Kronecker naturally followed Jacobi.

Remark 1: Niels Henrik Abel (5 August 1802 – 6 April 1829) was a Norwegian mathematician who, despite living in poor conditions and dying at the age of 26, made major contributions to mathematics. Abel was an innovator in the field of elliptic functions, discoverer of Abelian functions and a pioneer in the development of several branches of modern mathematics. (Wikipedia)

Remark 2: Carl Jacobi, in full Carl Gustav Jacob Jacobi (born December 10, 1804, Potsdam, Prussia [Germany] – died February 18, 1851, Berlin), was a German mathematician who, with Niels Henrik Abel of Norway, founded the theory of elliptic functions. (Encyclopædia Britannica)

Remark 3: Leopold Kronecker (born December 7, 1823 – died December 29, 1891) was a German mathematician who worked on number theory, algebra and elliptic functions.

1.2. *Complex multiplication in the sense of Abel*

In the last section §X of his paper [Ab-1827] mentioned above, Abel considered a problem on a separable differential equation of the form

$$\frac{dy}{\sqrt{(1-y^2)(1+\mu y^2)}} = \alpha \frac{dx}{\sqrt{(1-x^2)(1+\mu x^2)}},$$

with α, $\mu \in \mathbb{C}$ to solve it under the condition that the functions x and y have an algebraic relation $P(x,y) = 0$ with $P(X,Y) \in \mathbb{C}[X,Y]$; and he stated the two theorems:

Theorem I: If α is a real number, then it must be a rational number.

Theorem II: If α is a complex number [and not real], then it must be of the form $\alpha = m \pm \sqrt{-1} \cdot \sqrt{n},\ m, n \in \mathbb{Q},\ (n > 0)$. In the case, furthermore, the values of the parameter μ of the equation must satisfy a certain equation which has infinitely many real or imaginary roots. Conversely, each of such values of μ gives an algebraically solvable equation.

He also gave examples; namely,

$$(\alpha, \mu) = \left(\sqrt{-3},\ \ (2+\sqrt{3})^2\right), \qquad \left(\sqrt{-5},\ \left(2+\sqrt{5}+2\sqrt{2+\sqrt{5}}\right)^2\right).$$

In *Addition au mémoire précédent* attached at the end of the paper [Ab-1827], Abel gave a proof to Theorem I.

Then in his paper [Ab-1828b], Abel made a rigorous investigation of transformation of general elliptic differential forms which gives

$$\frac{dy}{\sqrt{(1-c_1^2y^2)(1-e_1^2y^2)}} = \pm a \frac{dx}{\sqrt{(1-c^2x^2)(1-e^2x^2)}},$$

where y is an algebraic function of x, either rational or irrational. At the end of this paper, he applied his results to the special case where $c_1 = c = 1$ and $e_1 = e$, and gave the proof to his Theorem II.

More precise results, especially, on the parameter μ of Theorem II will be presented later in §4.4 of this article.

It should also be pointed out that Abel made a big contribution to algebraic equations as is well known. He was strongly interested in algebraically solvable equations ([Ab-1826a]) and found some quite new such equations related to elliptic functions; he also found commutativity of Galois groups of a fundamental case of solvable equations, that is, now called Abelian equations whose Galois group is commutative ([Ab-1829a]). Under the influence of the cyclotomic theory of Gauss [Ga-1801], Kronecker was interested in these works of Abel, too, and first proposed Kronecker–Weber theorem on Abelian equations over the field $\mathbb{Q}$ of rational numbers ([Kr-1853]) and then dreamed 'Kronecker's Dream in young days' on such equations over imaginary quadratic number fields ([Kr-1880]). These research programs set by Kronecker finally bore fruits in 1920's. Takagi [Ta-1920a] established his congruence class field theory to show how Abelian equations over an algebraic number field are controlled by arithmetic of the base field, and then Artin [Ar-1927] proved his general reciprocity law which was a definite answer to the most basic problem of algebraic number theory since

Gauss [Ga-1801]. These two results are so closely related as to be called Takagi–Artin class field theory. (See [Mi-2011].)

2. Complex multiplication of elliptic functions

2.1. *The elliptic function and its periods*

Here we interpret the complex multiplication in the sense of Abel as the one of an elliptic function. See Definition below.

Fix a complex number w_0, and for each $w \in \mathbb{C}$ take a path $\gamma(w)$ from w_0 to w. Then the integral

$$z = z(w) := \int_{\gamma(w)} \frac{dw}{\sqrt{(1-w^2)(1+\mu w^2)}} \qquad (*)$$

along $\gamma(w)$ on the complex plane is well determined for such a path $\gamma(w)$ that it does not 'contain' any loops simply going around an odd number of the four roots of the polynomial $(1-w^2)(1+\mu w^2)$ in the denominator of the integral. This condition on the path comes from the square root in the world of complex numbers. Here we consider only such *admissible* paths (cf. §3.3 and §3.4).

Note, for example, $\sqrt{z}$ cannot be well defined around $z = 0 \in \mathbb{C}$ as a good continuous function; for that we need to cut a neighborhood of $z = 0$ along a slit starting at the point.

More precisely, let $\alpha_1 := 1, \alpha_2 := -1, \alpha_3 := \sqrt{-\mu}^{-1}, \alpha_4 := -\sqrt{-\mu}^{-1}$ be the four roots of $(1-w^2)(1+\mu w^2)$. Take closed paths γ_1 and γ_2 simply going around each of pairs of two roots (α_1, α_2) and (α_1, α_3), respectively, looking the corresponding pair of two points on the lefthand side. Put

$$\varpi_1 := \int_{\gamma_1} \frac{dw}{\sqrt{(1-w^2)(1+\mu w^2)}},$$
$$\varpi_2 := \int_{\gamma_2} \frac{dw}{\sqrt{(1-w^2)(1+\mu w^2)}},$$

and let $\mathfrak{m}$ be the submodule of the additive group $\mathbb{C}$ generated by ϖ_1 and ϖ_2, called the module of periods:

$$\mathfrak{m} := \mathbb{Z} \cdot \varpi_1 + \mathbb{Z} \cdot \varpi_2.$$

Proposition 1: For every closed admissible path γ, there is a pair $m, n \in \mathbb{Z}$ such that

$$\int_{\gamma} \frac{dw}{\sqrt{(1-w^2)(1+\mu w^2)}} = m\varpi_1 + n\varpi_2 \quad \in \mathfrak{m}.$$

Proposition 2: The module $\mathfrak{m}$ is of rank 2 over $\mathbb{Z}$. Moreover, the generators ϖ_1 and ϖ_2 are (linearly) independent over $\mathbb{R}$; that is, the quotient ϖ_1/ϖ_2 is not a real number.

Remark: Actually we have to consider the integrals on the Riemann sphere $\mathbb{C}^* = \mathbb{C} \cup \{\infty\}$. The differential form $dw/\sqrt{(1-w^2)(1+\mu w^2)}$ is holomorphic at $w = \infty \in \mathbb{C}^*$ as is easily seen by taking a local parameter $t = 1/w$. More exactly speaking, we have to work on the closed Riemann surface $\mathcal{R}$ of the differential form; it is a double covering of $\mathbb{C}^*$ which ramifies at four point $\alpha_j, j = 1, 2, 3, 4$; it is a model of the complete algebraic curve defined by $u^2 = (1-w^2)(1+\mu w^2)$ and is of genus 1. Then the form is holomorphic everywhere on $\mathcal{R}$. This is, however, too far advanced from the days of Abel.

Let us go back to the integral $(*)$. The value $z(w)$ is not uniquely determined by $w \in \mathbb{C}$, but so is the coset $z(w) + \mathfrak{m}$ of the quotient group $\mathbb{C}/\mathfrak{m}$. That is, the 'function' $z(w)$ is multi-valued. Hence the inverse function $w = w(z)$ is periodic with two independent periods ϖ_1 and ϖ_2: $w(z+\varpi_1) = w(z+\varpi_2) = w(z)$, and hence, $w(z+\varpi) = w(z)$ for $\varpi \in \mathfrak{m}$.

Since, moreover, the differential form $dw/\sqrt{(1-w^2)(1+\mu w^2)}$ is holomorphic at $w = \infty \in \mathbb{C}^*$, we also have a well determined coset $z(\infty) + \mathfrak{m}$ corresponding to $w = \infty$. This means that the function $w(z)$ has a pole at $z = z(\infty) + \varpi \in \mathbb{C}$ for every $\varpi \in \mathfrak{m}$.

Definition (Elliptic function): An *elliptic function* $f(z)$ is a meromorphic function on $\mathbb{C}$ which has two independent periods.

2.2. *Complex multiplication of an elliptic function*

It is now not difficult to see when an elliptic function has complex multiplication. Indeed, the above elliptic function $w(z)$ for the parameter μ has complex multiplication by α in the sense of Abel if $x = w(z)$ and $y = w(\alpha \cdot z)$ with $\alpha \in \mathbb{C} - \mathbb{R}$ have an algebraic relation $P(x, y) = 0$ with $P(X, Y) \in \mathbb{C}[X, Y]$. This observation allows us to give the following definition of complex multiplication of an elliptic function.

Definition (Complex multiplication of elliptic functions): An elliptic function $f(z)$ has *complex multiplication* by $\alpha \in \mathbb{C} - \mathbb{R}$ if $x = f(z)$ and $y = y(z) = f(\alpha \cdot z)$ have an algebraic relation $P(x, y) = 0$ with $P(X, Y) \in \mathbb{C}[X, Y]$.

The next step we take is to express the property of an elliptic function by means of the module of the periods.

Proposition: Let $f(z)$ be a non-constant elliptic function, and $\mathfrak{m}$ be the module of the periods of $f(z)$; that is

$$\mathfrak{m} = \{\varpi \in \mathbb{C} \mid f(z+\varpi) = f(z)\}.$$

Then for $\alpha \in \mathbb{C}$, two functions $x = f(z)$ and $y = f(\alpha \cdot z)$ have an algebraic relation $P(x,y) = 0$ with $P(X,Y) \in \mathbb{C}[X,Y]$ if and only if the submodule $\mathfrak{m} \cap \alpha \cdot \mathfrak{m}$ of the period module $\mathfrak{m}$ is of finite index.

Sketch of the proof

The module of the periods of $f(\alpha \cdot z)$ is $\alpha^{-1} \cdot \mathfrak{m} = \{\alpha^{-1}\varpi \mid \varpi \in \mathfrak{m}\}$. Multiplying by α, we see that the submodule $\mathfrak{m} \cap \alpha^{-1} \cdot \mathfrak{m}$ is of finite index in $\alpha^{-1} \cdot \mathfrak{m}$ if and only if $\mathfrak{m} \cap \alpha \cdot \mathfrak{m}$ is of finite index in $\mathfrak{m}$.

Suppose now that $P(x,y) = 0$ with $P(X,Y) \in \mathbb{C}[X,Y]$ for $x := f(z), y = y(z) := f(\alpha \cdot z)$. Then for each $\varpi \in \mathfrak{m}$ we have $P(f(z), y(z+\varpi)) = 0$ because $f(z+\varpi) = f(z)$. In other words, $y(z+\varpi)$ for $\varpi \in \mathfrak{m}$ is also a root of the polynomial $P(f(z), Y) \in \mathbb{C}(f(z))[Y]$ in Y. Since the set of all meromorphic functions on $\mathbb{C}$ form a (commutative) field, the number of the elements of the set $\{y(z+\varpi)) \mid \varpi \in \mathfrak{m}\}$ is finite and at most equal to the degree of the polynomial $P(f(z), Y)$ in Y. Since those elements of $\mathfrak{m}$ which trivially act on the set are nothing but ones in $\mathfrak{m} \cap \alpha^{-1} \cdot \mathfrak{m}$, this is a submodule of $\mathfrak{m}$ of finite index. Therefore it is of rank 2 as a $\mathbb{Z}$-module, and hence is of finite index in $\alpha^{-1} \cdot \mathfrak{m}$, too.

Conversely, suppose that $\mathfrak{m} \cap \alpha \cdot \mathfrak{m}$ is of finite index in $\mathfrak{m}$. Then $\mathfrak{m} \cap \alpha^{-1} \cdot \mathfrak{m}$ is of finite index in $\alpha^{-1} \cdot \mathfrak{m}$, and so, is of rank 2 as a $\mathbb{Z}$-module. Hence it is also of finite index in $\mathfrak{m}$. Take a set of representatives $\varpi_1, \ldots, \varpi_n \in \mathfrak{m}$ for the quotient group $\mathfrak{m}/\mathfrak{m} \cap \alpha^{-1} \cdot \mathfrak{m}$ and put

$$Q(Y) := \prod_{j=1}^{n} (Y - y(z+\varpi_j)).$$

Then the coefficients of $Q(Y)$ are meromorphic functions which admit all of the elements of $\mathfrak{m}$ as periods. It is well known that all the meromorphic functions (including constant functions) which admit all of the elements of $\mathfrak{m}$ as periods form a finitely generated *elliptic function field* $\mathcal{K}_{\mathfrak{m}}$ over $\mathbb{C}$. (See Theorem (Weierstrass $\wp$ function) in the next section.) Therefore $\mathcal{K}_{\mathfrak{m}}$ is a finite algebraic extension of the subfield $\mathbb{C}(f(z))$. Let $R(Y) := N_{\mathcal{K}_{\mathfrak{m}}/\mathbb{C}(f(z))}(Q(Y))$ be the norm of $Q(Y)$, that is, the product of all of the algebraic conjugates of

$Q(Y) \in \mathcal{K}_{\mathfrak{m}}(Y)$ over $\mathbb{C}(f(z))(Y)$. Then the coefficients of $R(Y)$ are functions in $\mathbb{C}(f(z))$. Multiply $R(Y)$ by the least common multiple of the denominators of the coefficients and replace $f(z)$ by X to get $P(X,Y) \in \mathbb{C}[X,Y]$ with the property $P(f(z), y(z)) = P(f(z), f(\alpha \cdot z)) = 0$.

2.3. *The elliptic function field determined by a module of periods*

Suppose that a $\mathbb{Z}$-module $\mathfrak{m} = \mathbb{Z} \cdot \varpi_1 + \mathbb{Z} \cdot \varpi_2 \subset \mathbb{C}$ is of rank 2 and that $\tau := \varpi_1/\varpi_2$ is not a real number. (We may assume that the imaginary part of τ is positive; if not, change the two indices to get it.) Then there always exist non-trivial elliptic functions which admit $\mathfrak{m}$ as their module of the periods. A pair of the most convenient ones for us may be the Weierstrass $\wp$ function and its derivative function $\wp'$. The infinite series of rational functions of z,

$$\wp(z) = \wp(\mathfrak{m}; z) := \frac{1}{z^2} + \sum_{\varpi \in \mathfrak{m} - \{0\}} \left\{ \frac{1}{(z-\varpi)^2} - \frac{1}{\varpi^2} \right\}$$

converges on $\mathbb{C} - \mathfrak{m}$ uniformly in the wide sense and gives a meromorphic function on $\mathbb{C}$ with poles of order 2 at each $z = \varpi \in \mathfrak{m}$. Therefore, its derived function

$$\wp'(z) = \wp'(\mathfrak{m}; z) = \frac{d\wp(\mathfrak{m}; z)}{dz} = -2 \sum_{\varpi \in \mathfrak{m}} \frac{1}{(z-\varpi)^3}$$

is also meromorphic on $\mathbb{C}$ with poles of order 3 at each $z = \varpi \in \mathfrak{m}$. (See, for example, Lang [La-1987, Ch. 1].)

Proofs of the facts on Weierstrass $\wp$ function stated in the next theorem will also be found in [La-1987, Ch. 1].

Theorem (Weierstrass $\wp$ function): The notation and the assumptions being as above, $\wp(z)$ and $\wp'(z)$ are elliptic functions whose module of periods coincide with $\mathfrak{m}$. They satisfy *Weierstrass' equation*

$$\wp'(z)^2 = 4\wp(z)^3 - g_2\wp(z) - g_3,$$

$$g_2 = g_2(\mathfrak{m}) := 60 \sum_{\varpi \in \mathfrak{m} - \{0\}} \frac{1}{\varpi^4}, \qquad (\dagger)$$

$$g_3 = g_3(\mathfrak{m}) := 140 \sum_{\varpi \in \mathfrak{m} - \{0\}} \frac{1}{\varpi^6}.$$

The existence of the elliptic function $\wp(z) = \wp(\mathfrak{m}; z)$ implies that the cubic polynomial on the right hand side of the equation does not have any multiple roots, and hence its discriminant is not equal to 0: $g_2^3 - 27g_3^2 \neq 0$. Furthermore, $\wp(z)$ and $\wp'(z)$ generate the elliptic function field $\mathcal{K}_\mathfrak{m}$: $\mathcal{K}_\mathfrak{m} = \mathbb{C}(\wp(z), \wp'(z))$.

2.4. *Reasoning of Abel's two theorems*

Here we demonstrate Abel's Theorem I and the first half of Theorem II. A precise mathematical statement for the latter half of Theorem II on the parameter μ will be given later in §4.4.

Let the submodule $\mathfrak{m}$ of $\mathbb{C}$ be as above. Assume that $\mathfrak{m} \cap \alpha \cdot \mathfrak{m}$ is of finite index in $\mathfrak{m}$ for $\alpha \in \mathbb{C}$. Then the $\mathbb{Z}$-module $\mathfrak{m} \cap \alpha \cdot \mathfrak{m}$ is also of rank 2, and hence, of finite index in $\alpha \cdot \mathfrak{m}$, too. There exists, therefore, a positive integer N such that

$$N(\alpha \cdot \mathfrak{m}) = (N\alpha) \cdot \mathfrak{m} \subset \mathfrak{m}.$$

If we express the inclusion relation with the generators ϖ_1 and ϖ_2 of $\mathfrak{m}$, we find $a, b, c, d \in \mathbb{Z}$ which give the equation

$$N\alpha \cdot \begin{pmatrix} \varpi_1 \\ \varpi_2 \end{pmatrix} = \begin{pmatrix} a & b \\ c & d \end{pmatrix} \begin{pmatrix} \varpi_1 \\ \varpi_2 \end{pmatrix}.$$

Multiplying both sides by ϖ_2^{-1} we get

$$N\alpha \cdot \begin{pmatrix} \tau \\ 1 \end{pmatrix} = \begin{pmatrix} a & b \\ c & d \end{pmatrix} \begin{pmatrix} \tau \\ 1 \end{pmatrix},$$

and hence $N\alpha = c\tau + d$. (1) If $c = 0$, then α is a rational number. (2) if $c \neq 0$, then α must be imaginary because so is τ. Moreover, $N\alpha$ is an eigenvalue of the square matrix, and so, a root of the quadratic polynomial $X^2 - (a+d)X + ad - bc$ with $-(a+d),\ ad - bc \in \mathbb{Q}$. Hence, α and τ generate the same imaginary quadratic field $\mathbb{Q}(\alpha) = \mathbb{Q}(\tau)$ in this case. This shows Abel's Theorem I and the former half of Theorem II.

Remark: If τ is, conversely, an imaginary quadratic number whose imaginary part is positive, then $\varpi_1 := \tau$ and $\varpi_2 := 1$ give the module of periods $\mathfrak{m} := \mathbb{Z} \cdot \varpi_1 + \mathbb{Z} \cdot \varpi_2$ of $\wp$ function $\wp(\mathfrak{m}; z)$. Since $\tau^2 - a\tau + b = 0$ for some $a, b \in \mathbb{Q}$, it is not hard to see that $\alpha := \tau$ gives complex multiplication of $\wp(\mathfrak{m}; z)$. Indeed, let N be the least common multiple of the denominators of a and b. Then we have $N(\alpha \cdot \mathfrak{m}) \subset \mathfrak{m}$ because $N\alpha \cdot \tau = N\tau^2 = -(Na)\tau - (Nb) \in \mathfrak{m}$.

Here it makes the world of elliptic functions very clear that there exist those non-trivial elliptic functions which admit an arbitrarily given $\mathbb{Z}$-module of rank 2 as their periods. There might, however, not always exist so many independent meromorphic functions on a higher dimensional complex torus as its dimension (see §6.1).

3. Geometry of elliptic functions

3.1. *The elliptic curve associated to a complex torus*

As in the preceding section, we consider the $\mathbb{Z}$-module $\mathfrak{m} = \mathbb{Z} \cdot \varpi_1 + \mathbb{Z} \cdot \varpi_2$ of rank 2, the module of periods. Since $\mathfrak{m}$ is a discrete and closed subgroup of the additive group $\mathbb{C}$, the complex torus $\mathbb{C}/\mathfrak{m}$ is a compact Riemann surface of genus 1. The field of all meromorphic functions on it may be identified with the field $\mathcal{K}_{\mathfrak{m}}$ of all of those elliptic functions which admit $\mathfrak{m}$ as their periods. The generators $\wp(z) := \wp(\mathfrak{m}; z)$ and its derived function $\wp'(z) = \wp'(\mathfrak{m}; z)$ define an embedding of $\mathbb{C}/\mathfrak{m}$ into the projective plane $\mathbb{P}^2(u : v : w)$ with a projective coordinate system $(u : v : w)$ through $\Phi : \mathbb{C} \to \mathbb{P}^2(\mathbb{C})$ defined as follows:

$$\Phi(z) = \begin{cases} (\wp(z) : \wp'(z) : 1) & z \in \mathbb{C} - \mathfrak{m}, \\ P_\infty = (0 : 1 : 0) & z \in \mathfrak{m}. \end{cases}$$

Then by Weierstrass' equation (†), the image of $\mathbb{C}/\mathfrak{m}$ is the *elliptic curve* $E_{\mathfrak{m}}$ defined by the homogeneous equation

$$E_{\mathfrak{m}} : \qquad v^2 w = 4u^3 - g_2 u w^2 - g_3 w^3$$

where g_2, g_3 are Eisenstein series defined in (†). Actually, Φ induces an isomorphism of $\mathbb{C}/\mathfrak{m}$ onto $E_{\mathfrak{m}}$ over $\mathbb{C}$ which maps the origin 0 to the unique point $P_\infty = (0 : 1 : 0)$ of $E_{\mathfrak{m}}$ at infinity. Furthermore, the natural additive group structure of the complex torus $\mathbb{C}/\mathfrak{m}$ implies the one on the elliptic curve $E_{\mathfrak{m}}$ which is a geometric reflection of the additive formulas of the elliptic functions $\wp(\mathfrak{m}; z)$ and $\wp'(\mathfrak{m}; z)$.

3.2. *Complex multiplication on an elliptic curve*

The notation and the assumptions being as in §3.1, suppose that $\mathfrak{m} \cap \alpha \cdot \mathfrak{m}$ is of finite index in $\mathfrak{m}$ for $\alpha \in \mathbb{C}$. Take such a positive integer N as we have

$(N\alpha) \cdot \mathfrak{m} \subset \mathfrak{m}$. Then the analytic automorphism of the additive group $\mathbb{C}$ obtained by multiplication of $N\alpha$ induces an analytic endomorphism of the additive group $\mathbb{C}/\mathfrak{m}$.

Proposition 1: Let $\mathfrak{m}'$ be another $\mathbb{Z}$-module of the same kind as $\mathfrak{m}$ and suppose that $\varphi : \mathbb{C}/\mathfrak{m} \to \mathbb{C}/\mathfrak{m}'$ is an analytic homomorphism of additive groups. Then there is such a complex number a as we have $\varphi(z \bmod \mathfrak{m}) = az \bmod \mathfrak{m}'$ for $z \in \mathbb{C}$. In particular, we have the inclusion relation $a \cdot \mathfrak{m} \subset \mathfrak{m}'$.

For a proof, see Birkenhake and Lange [BL-2010, Ch. 1, §1.2] or Lang [La-1986, Ch. 1].

This proposition implies, in particular, that every pair of endomorphisms of $\mathbb{C}/\mathfrak{m}$ is commutative; and hence the observation in §2.4 easily implies the next proposition.

Proposition 2: Let $\mathrm{End}(\mathbb{C}/\mathfrak{m})$ be the ring of all analytic endomorphisms of $\mathbb{C}/\mathfrak{m}$.
(1) The *endomorpism algebra* $\mathrm{End}_{\mathbb{Q}}(\mathbb{C}/\mathfrak{m}) := \mathrm{End}(\mathbb{C}/\mathfrak{m}) \otimes_{\mathbb{Z}} \mathbb{Q}$ is isomorphic either to $\mathbb{Q}$ or to an imaginary quadratic field.
(2) For $\alpha \in \mathbb{C}$, $\mathfrak{m} \cap \alpha \cdot \mathfrak{m}$ is of finite index in $\mathfrak{m}$ if and only if $\mathrm{End}_{\mathbb{Q}}(\mathbb{C}/\mathfrak{m})$ contains a field which is isomorphic to the number field $\mathbb{Q}(\alpha)$. Furthermore, $\mathbb{Q}(\alpha)$ is equal either to $\mathbb{Q}$ or to $\mathbb{Q}(\tau)$, $\tau = \varpi_1/\varpi_2$. The latter case occurs if and only if α is an imaginary quadratic number.

Definition (Complex multiplication on elliptic curves): An elliptic curve E is said to have *complex multiplication* if its endomorphism algebra $\mathrm{End}_{\mathbb{Q}}(E)$ is isomorphic to an imaginary quadratic number field.

3.3. *The Riemann surface of an elliptic integral*

Let us now study the Riemann surface $\mathcal{R}$ of an elliptic differential form $dw/\sqrt{P(w)}$ where $P(X)$ is a polynomial of degree 4 which does not have any multiple roots. As we indicated in Remark in §2.1 $\mathcal{R}$ is a double covering of the Riemann sphere $\mathbb{C}^* := \mathbb{C} \cup \{\infty\}$ which ramifies at the distinct four roots $\alpha_1, \alpha_2, \alpha_3, \alpha_4$ of $P(X)$. Hence we may regard it as the elliptic curve on the projective plane whose affine form is defined by

$$E : \qquad y^2 = P(x).$$

It is of genus 1 and has just one point $P_\infty = (0 : 1 : 0)$ at infinity, and hence, is an elliptic curve defined over the number field generated by the coefficients of $P(X)$ over $\mathbb{Q}$. The differential form $dw/\sqrt{P(w)}$ gives a holomorphic 1-form on $\mathcal{R}$. Note that we may take a local parameter around the point on $\mathcal{R}$ over each root $\alpha_j, (1 \leq j \leq 4)$, of $P(w)$ as $t := \sqrt{w - \alpha_j}$.

Let γ_1' and γ_2', respectively, be simple closed curves on $\mathbb{C}^*$ which simply go around each of pairs of points $\{\alpha_1, \alpha_2\}$ and $\{\alpha_1, \alpha_3\}$, and let γ_1 and γ_2, respectively, be the ones on $\mathcal{R}$ over γ_1' and γ_2'. Then γ_1 and γ_2 give a couple of generators of the homology group $\mathrm{H}_1(\mathcal{R}, \mathbb{Z})$ which is a free $\mathbb{Z}$-module of rank 2. Put

$$\varpi_1 := \int_{\gamma_1} \frac{dw}{\sqrt{P(w)}}, \qquad \varpi_2 := \int_{\gamma_2} \frac{dw}{\sqrt{P(w)}},$$

$$\mathfrak{m} := \mathbb{Z} \cdot \varpi_1 + \mathbb{Z} \cdot \varpi_2.$$

Then $\mathfrak{m}$ is a $\mathbb{Z}$-module of rank 2, and $\tau := \varpi_1/\varpi_2$ is not a real number; we may assume that the imaginary part of τ is positive as usual.

3.4. *The relation of the Riemann surface with the complex torus*

The notation and the assumptions being as above, we now establish a canonical isomorphism from the Riemann surface $\mathcal{R}$ onto the complex torus $\mathbb{C}/\mathfrak{m}$. This is a kind of geometric reasoning of the former half of §2.1.

Fix a base point $P_0 \in \mathcal{R}$. For each point $P \in \mathcal{R}$, take a path $\gamma(P)$ from P_0 to P on $\mathcal{R}$. Since the differential form $dw/\sqrt{P(w)}$ is holomorphic everywhere on $\mathcal{R}$, the value $\psi(\gamma(P)) := \int_{\gamma(P)} (1/\sqrt{P(z)})\, dw$ is well determined. The value is not, however, well determined only by the point P. Indeed, let $\gamma'(P)$ be another path from P_0 to P on $\mathcal{R}$. Then the closed path $\gamma(P) - \gamma'(P)$ is homologous to a linear combination $m\gamma_1 + n\gamma_2$ with some pair $m, n \in \mathbb{Z}$ on $\mathcal{R}$. Therefore, the coset $\Psi(P) := \psi(\gamma(P)) \bmod \mathfrak{m}$ of $\mathbb{C}/\mathfrak{m}$ is well determined by the point $P \in \mathcal{R}$; and hence, a map $\Psi : \mathcal{R} \to \mathbb{C}/\mathfrak{m}$ is now established.

Proposition: The map $\Psi : \mathcal{R} \to \mathbb{C}/\mathfrak{m}$ defined above from the compact Riemann surface $\mathcal{R}$ to the complex torus is an analytic isomorphism.

Here, it should be noted that in the days of Abel the geometry briefly explained in this section was not ready in use. On the contrary, it was Abel's works on elliptic functions and on Abelian integrals that inspired Jacobi, Riemann, Kronecker, Weierstrass, and so on, to develop mathematics which we now enjoy.

Remark: In the preceding section, we started with a polynomial $P(X)$ of degree 4 which does not have any multiple roots. We may also proceed our argument with a polynomial $P(X)$ of degree 3 without any multiple roots. In this case the Riemann surface $\mathcal{R}$ for the form $dw/\sqrt{P(w)}$ is a double covering of the Riemann sphere $\mathbb{C}^*$ which ramifies not only at the three roots $\alpha_1, \alpha_2, \alpha_3$ of $P(X)$ but also at the point $\alpha_4 := \infty \in \mathbb{C}^*$. (Hence especially, we may take $t := 1/\sqrt{w}$ as a local parameter at the point on $\mathcal{R}$ over $w = \infty$.)

4. The elliptic modular function and singular moduli

In this section, we give an explicit mathematical statement for the latter half of Abel's Theorem II in §1.2.

4.1. *The elliptic modular function*

Suppose that the $\mathbb{Z}$-module $\mathfrak{m} = \mathbb{Z}\cdot\varpi_1 + \mathbb{Z}\cdot\varpi_2 \subset \mathbb{C}$ is of rank 2 and the imaginary part of $\tau = \varpi_1/\varpi_2$ is positive. Then $\mathfrak{m}$ determines an elliptic function field $\mathcal{K}_\mathfrak{m}$; it consists of all elliptic functions (including constant functions) whose modules of periods contain $\mathfrak{m}$. Then $\mathcal{K}_\mathfrak{m} = \mathbb{C}(\wp(\mathfrak{m}; z), \wp'(\mathfrak{m}; z))$ where $\wp(\mathfrak{m}; z)$ and $\wp'(\mathfrak{m}; z)$ are Weierstrass $\wp$ function and its derived function, respectively, as they were explicitly introduced in §2.3 above. They satisfy Weierstrass' equation (†). Note that $g_2^3 - 27g_3^2 \neq 0$.

Suppose, furthermore, that an abstract function field $\mathbb{C}(\xi, \eta)$ with a relation

$$\eta^2 = 4\xi^3 - \gamma_2\xi - \gamma_3,$$
$$\gamma_2^3 - 27\gamma_3^2 \neq 0, \quad \gamma_2,\ \gamma_3\ \in\ \mathbb{C},$$

is given. Then we have the following theorem. (For a proof, see Lang [La-1987, Ch. 3], for example.)

Theorem (Moduli of Elliptic Function Fields): Two fields $\mathcal{K}_{\mathfrak{m}}$ and $\mathbb{C}(\xi, \eta)$ are isomorphic over $\mathbb{C}$ as abstract fields if and only if

$$\frac{g_2^3}{g_2^3 - 27g_3^2} = \frac{\gamma_2^3}{\gamma_2^3 - 27\gamma_3^2}.$$

The left-hand side of the equation of the theorem is determined by $\mathfrak{m}$; it is easy to see that the value stays unchanged if $\mathfrak{m}$ is replaced by $\lambda \cdot \mathfrak{m}$ with $\lambda \in \mathbb{C} - \{0\}$. Since the module $\varpi_2^{-1} \cdot \mathfrak{m}$ is generated by $\tau = \varpi_1/\varpi_2$ and 1 over $\mathbb{Z}$, the value is determined by $\tau \in \mathbb{C}, \Im(\tau) > 0$; hence now, put

$$J(\tau) = J(\mathfrak{m}) := \frac{12^3 g_2(\mathfrak{m})^3}{g_2(\mathfrak{m})^3 - 27g_3(\mathfrak{m})^2}.$$

This is a holomorphic function on the upper half complex plane

$$\mathfrak{H} := \{\tau \in \mathbb{C} \mid \Im(\tau) > 0\},$$

and is called the *elliptic modular function.* (The *j-invariant* of the elliptic curve defined by the above Weierstrass' equation (†) is $J(\tau)/12^3$. The constant 12^3 for $J(\tau)$ is traditional and simplifies certain analytic expressions of the function.)

Remark 1: Let $\mathfrak{m}'$ be another $\mathbb{Z}$-module of the same kind as $\mathfrak{m}$. Since $\mathcal{K}_{\mathfrak{m}}$ is the field of all meromorphic functions on the complex torus $\mathbb{C}/\mathfrak{m}$ and so is $\mathcal{K}_{\mathfrak{m}'}$ for $\mathbb{C}/\mathfrak{m}'$, an isomorphism of the two tori implies an isomorphism of the two fields, and indeed, vice versa. The theorem states, however, an analytic fact which is quite different in its nature from the geometric fact on isomorphisms of complex tori which is given by Proposition 1 in §3.2. This suggests that we may, indeed, have a deep result from the two facts.

Remark 2: For an arbitrarily given pair of complex numbers $\gamma_2, \gamma_3 \in \mathbb{C}$, suppose that $\gamma_2^3 - 27\gamma_3^2 \neq 0$. Then there always exists such a $\mathbb{Z}$-module $\mathfrak{m} \subset \mathbb{C}$ of rank 2 with $\mathbb{R} \cdot \mathfrak{m} = \mathbb{C}$ as we have $\gamma_2 = g_2(\mathfrak{m})$ and $\gamma_3 = g_3(\mathfrak{m})$. (Cf. Lang [La-1987, Ch. 3, §3], for example.) Hence an abstract elliptic function field $\mathbb{C}(\xi, \eta)$ defined by a Weierstrass form $\eta^2 = 4\xi^3 - \gamma_2\xi - \gamma_3$, $\gamma_2, \gamma_3 \in \mathbb{C}, \gamma_2^3 - 27\gamma_3^2 \neq 0$, is always realized as $\mathcal{K}_{\mathfrak{m}} = \mathbb{C}(\wp(\mathfrak{m}; z), \wp'(\mathfrak{m}; z))$ for a suitable $\mathbb{Z}$-module $\mathfrak{m}$.

4.2. *Singular moduli*

Here in this subsection, we consider a $\mathbb{Z}$-module of the form

$$\mathfrak{m} = \mathbb{Z} \cdot \tau + \mathbb{Z} \cdot 1, \qquad \tau \in \mathfrak{H}.$$

Furthermore, we confine ourselves to the case where the complex torus and hence the elliptic functions associated with τ admit complex multiplication.

Assumption: The number field $\mathbb{Q}(\tau)$ for $\tau \in \mathfrak{H}$ is assumed to be an imaginary quadratic field.

Then the above module $\mathfrak{m}$ is a submodule of the additive group of the imaginary quadratic field $\mathbb{Q}(\tau)$.

Definition (Singular moduli): The value $J(\tau)$, for $\tau \in \mathfrak{H}$ satisfying the assumption, is called a *singular modulus.*

We have to replace the role of the parameter μ of Abel's differential equation in the latter half of Abel's Theorem II in §1.2 by the corresponding singular modulus at α to give an explicit mathematical statement.

4.3. *Arithmetic of imaginary quadratic number fields*

Let the assumption on $\tau \in \mathfrak{H}$ be as in the preceding section. We now consider all $\mathbb{Z}$-modules of rank 2 in the quadratic field $F := \mathbb{Q}(\tau)$, and introduce hierarchy among such $\mathbb{Z}$-modules. Let $\mathfrak{a}$ be one of them.

Definition: The subring $\mathcal{O}(\mathfrak{a}) := \{\alpha \in F \mid \alpha \cdot \mathfrak{a} \subset \mathfrak{a}\}$ of the quadratic field F is called the *order* of $\mathfrak{a}$ (after Dedekind).

Then $\mathfrak{a}$ is a 'fractional ideal' of its order $\mathcal{O}(\mathfrak{a})$; that is, $N \cdot \mathfrak{a} \subset \mathcal{O}(\mathfrak{a})$ is an ideal of the ring $\mathcal{O}(\mathfrak{a})$ for some positive integer N.

There is the maximal order $\mathcal{O}_F$ among orders that contains all the others: namely,

$$\mathcal{O}_F = \begin{cases} \mathbb{Z} \cdot 1 + \mathbb{Z} \cdot \frac{1+\sqrt{d}}{2} & \text{if } d \equiv 1 \mod 4, \\ \mathbb{Z} \cdot 1 + \mathbb{Z} \cdot \sqrt{d} & \text{if } d \not\equiv 1 \mod 4, \end{cases}$$

where $d \in \mathbb{Z}, d < 0$, is square-free and $\mathbb{Q}(\sqrt{d}) = F$. (This is the ring of all algebraic integers in F.)

Restriction: Hereafter in this chapter, we consider just the most fundamental case where the order of the module $\mathfrak{a}$ is maximal. Therefore, we restrict ourselves to so called fractional ideals of the quadratic field F as our modules.

4.4. *The ideal class group and the Hilbert class field*

The notation and the assumptions being as above, let us denote the multiplicative group of all fractional ideals of the quadratic field $F = \mathbb{Q}(\sqrt{d}) = \mathbb{Q}(\tau)$ by $\mathcal{I}_F$, and the subgroup of $\mathcal{I}_F$ consisting of all principal ideals by $\mathcal{P}_F$: $\mathcal{P}_F = \{\alpha\mathcal{O}_F \mid \alpha \in F, \alpha \neq 0\}$.

For $\mathfrak{a}, \mathfrak{b} \in \mathcal{I}_F$, the complex tori $\mathbb{C}/\mathfrak{a}$ and $\mathbb{C}/\mathfrak{b}$ are isomorphic if and only if there is such $\alpha \in \mathbb{C}$ as we have $\alpha \cdot \mathfrak{a} = \mathfrak{b}$; that is, the cosets of $\mathfrak{a}$ and $\mathfrak{b}$ in the quotient group $\mathcal{I}_F/\mathcal{P}_F$ coincide because the equality implies $\alpha \in F, \alpha \neq 0$. The quotient group $\mathrm{Cl}_F := \mathcal{I}_F/\mathcal{P}_F$ is the *ideal class group* of F, which has a quite different origin closely related to the equivalence classes of those binary quadratic forms with coefficients in $\mathbb{Z}$ the discriminants of which coincide with d or $4d$ according as either $d \equiv 1 \mod 4$ or $d \not\equiv 1 \mod 4$. It is a finite Abelian group; its order h_F is the class number of F.

For an ideal $\mathfrak{a} \in \mathcal{I}_F$, here we may denote the singular modulus associated with the module $\mathfrak{a}$ by $J(\mathfrak{a})$.

Theorem (Singular moduli): (1) For an ideal $\mathfrak{a} \in \mathcal{I}_F$, the value of the singular modulus $J(\mathfrak{a})$ depends only on the coset of $\mathfrak{a}$ in the class group Cl_F.
(2) Let $\mathfrak{m}_1, \mathfrak{m}_2, \ldots, \mathfrak{m}_{h_F} \in \mathcal{I}_F$ be a set of representatives of the quotient group $\mathrm{Cl}_F = \mathcal{I}_F/\mathcal{P}_F$. Then the translation by $\mathfrak{a}$ induced from $\mathfrak{m} \mapsto \mathfrak{a}\mathfrak{m}$ on Cl_F gives a permutation on the set of the singular moduli $\{J(\mathfrak{m}_1), J(\mathfrak{m}_2), \ldots, J(\mathfrak{m}_{h_F})\}$.
(3) The coefficients of the polynomial

$$P(X) := (X - J(\mathfrak{m}_1))(X - J(\mathfrak{m}_2)) \cdots (X - J(\mathfrak{m}_{h_F}))$$

belong to F, and $P(X)$ is irreducible over F.
(4) Each $J(\mathfrak{m}_j)$, $(1 \leq j \leq h_F)$, generates the same Galois extension $\tilde{F}/F$, and the Galois group $\mathrm{Gal}(\tilde{F}/F)$ is an Abelian group which is isomorphic to the class group Cl_F. Furthermore, the extension $\tilde{F}$ of F is the *Hilbert class field* of F; that is, the maximal unramified Abelian extension of F.
(5) (Reciprocity Law for Elliptic Modular Function) For a prime ideal $\mathfrak{p}$ of F, let $\sigma(\mathfrak{p})$ be the Frobenius automorphism of $\mathfrak{p}$ in $\mathrm{Gal}(\tilde{F}/F)$. Then we have $J(\mathfrak{a})^{\sigma(\mathfrak{p})} = J(\mathfrak{p}^{-1}\mathfrak{a})$ for every fractional ideal $\mathfrak{a}$ of F.

Remark: Shimura's book [Sh-1971] contains a fully developed modern results on complex multiplication of elliptic functions and elliptic modular functions.

5. The case of hyperelliptic integrals

We follow a way from complex multiplication on elliptic curves to that on Abelian varieties, via Jacobian varieties of hyperelliptic curves. A more precise sketch will be seen in Stepanov [St-1994, Ch. 7, Exercises, No. 3*]. One of the basic references on the subject is Lang [La-1972]. For our approach over the complex number field $\mathbb{C}$, Birkenhake and Lange [BL-2010, Ch. 11] may also be helpful.

On hyperelliptic Abelian integrals, two papers of Abel were published before his death:

[Ab-1828c] Remarques sur quelques propriétés générales d'un certaine sorte de fonctions transcendantes, *Jour. reine angew. Math.* **Bd.3** (1828), 313–323;

[Ab-1829b] Démonstration d'une propriété d'une certaine classe de fonctions transcendantes, *Jour. reine angew. Math.* **Bd.4** (1829), 200–201.

He started, however, his investigations on Abelian integrals as early as in 1826; see

[Ab-1826b] Mémoire sur une propriété générale d'une classe trés étendue de fonctions transcendantes, *Œuvres complés de Niels Henrik Abel, Nouvelle Édition* I, 1881, pp.145–211.

This paper was submitted to Paris Academy in 1826. It seemed lost, however, and was later found on Cauchy's desk after his death to be published in 1841.

5.1. *Jacobi inversion problem*

After the death of Abel, pioneers like Jacobi, Riemann and others explored the ways to generalize elliptic functions on the basis of hyperelliptic integrals. It was Jacobi who found functions of two complex variables with four independent periods (Jacobi [Ja-1835]).

Theorem: Let $P(X)$ be a polynomial of degree 6 with no multiple roots. Put

$$z_1 := \int^{w_1} \frac{dw_1}{\sqrt{P(w_1)}} + \int^{w_2} \frac{dw_2}{\sqrt{P(w_2)}}, \quad z_2 := \int^{w_1} \frac{w_1\, dw_1}{\sqrt{P(w_1)}} + \int^{w_2} \frac{w_2\, dw_2}{\sqrt{P(w_2)}}.$$

Then the symmetric functions

$$s_1(z_1, z_2) := w_1 + w_2, \qquad s_2(z_1, z_2) := w_1 \cdot w_2$$

are well defined functions of two variables z_1 and z_2 with independent four periods on $\mathbb{C}^2$.

Then in 1846, he proposed 'Jacobi inversion problem' to develop a theory of Abelian functions for higher genus in [Ja-1846].

Remark: Carl Gustav Jacob Jacobi (10 December 1804 – 18 February 1851) was a German mathematician, who made fundamental contributions to elliptic functions, dynamics, differential equations, and number theory. His name is occasionally written as Carolus Gustavus Iacobus Iacobi in his Latin books, and his first name is sometimes given as Karl. Jacobi was the first Jewish mathematician to be appointed professor at a German university. (Wikipedia)

Note that the Riemann surface $\mathcal{R}$ of the above two differential forms $dw/\sqrt{P(w)}$ and $w\, dw/\sqrt{P(w)}$ is of genus 2, and that the two forms span the space of holomorphic 1-forms on $\mathcal{R}$.

5.2. *Hyperelliptic curves and their Jacobian varieties*

Suppose now that $P(X)$ is a polynomial of degree $2g+2, g \in \mathbb{Z}, g > 0$, without any multiple roots. Then the Riemann surface $\mathcal{R}_g$ of the differential form $dw/\sqrt{P(w)}$ is a *hyperelliptic curve* of genus g defined by the affine form

$$E_g : \qquad y^2 = P(x).$$

(Actually, $\mathcal{R}_g$ and E_g are determined by the polynomial $P(X)$.) The space $\mathrm{H}^{(1.0)}(\mathcal{R}_g)$ of all holomorphic 1-forms on $\mathcal{R}_g$ is spanned over $\mathbb{C}$ by the g

forms

$$\frac{dw}{\sqrt{P(w)}}, \quad \frac{w\,dw}{\sqrt{P(w)}}, \quad \ldots, \quad \frac{w^{g-1}\,dw}{\sqrt{P(w)}}.$$

Let us denote the dual space of the vector space $\mathrm{H}^{(1.0)}(\mathcal{R}_g)$ over $\mathbb{C}$ by $\hat{\mathrm{H}}^{(1.0)}(\mathcal{R}_g)$: $\hat{\mathrm{H}}^{(1.0)}(\mathcal{R}_g) := \mathrm{Hom}_{\mathbb{C}}(\mathrm{H}^{(1.0)}(\mathcal{R}_g), \mathbb{C})$. Then it is a vector space over $\mathbb{C}$ of dimension g.

Since the genus of $\mathcal{R}_g$ is equal to g, its homology group $\mathrm{H}_1(\mathcal{R}_g, \mathbb{Z})$ is a $\mathbb{Z}$-module of rank $2g$. Let $\gamma_1, \gamma_2, \ldots, \gamma_{2g-1}, \gamma_{2g}$ be simple closed paths on $\mathcal{R}_g$ which represent a system of generators of $\mathrm{H}_1(\mathcal{R}_g, \mathbb{Z})$ over $\mathbb{Z}$.

Now a path γ on $\mathcal{R}_g$ defines an element $\hat{\gamma}$ of $\hat{\mathrm{H}}^{(1.0)}(\mathcal{R}_g)$ by assigning the value of the integral $\hat{\gamma}(\omega) := \int_\gamma \omega$ to a form $\omega \in \mathrm{H}^{(1.0)}(\mathcal{R}_g)$. Hence we obtain a $\mathbb{Z}$-module Π_{2g} of rank $2g$ in $\hat{\mathrm{H}}^{(1.0)}(\mathcal{R}_g) \simeq \mathbb{C}^g$,

$$\Pi_{2g} := \mathbb{Z} \cdot \hat{\gamma}_1 + \mathbb{Z} \cdot \hat{\gamma}_2 + \cdots + \mathbb{Z} \cdot \hat{\gamma}_{2g},$$

where $\hat{\gamma}_j, 1 \leq j \leq 2g$, is the element of $\hat{\mathrm{H}}^{(1.0)}(\mathcal{R}_g)$ defined correspondingly by the closed path γ_j on $\mathcal{R}_g$.

Definition (Jacobian variety): The g-dimensional complex torus $\mathrm{Jac}(\mathcal{R}_g) := \hat{\mathrm{H}}^{(1.0)}(\mathcal{R}_g)/\Pi_{2g}$ of dimension g over $\mathbb{C}$ is called the *Jacobian variety* of the hyperelliptic curve $\mathcal{R}_g$.

Theorem (Jacobian variety): There exist g independent meromorphic functions on the Jacobian variety of $\mathcal{R}_g$. Furthermore, there is a canonical embedding of $\mathcal{R}_g$ into its Jacobian variety $\mathrm{Jac}(\mathcal{R}_g)$.

Actually, the Jacobian variety is an Abelian variety; cf. the proposition in §6.1 below. As for the latter half, indeed, fix a base point $P_0 \in \mathcal{R}_g$, and for each point $P \in \mathcal{R}_g$, take a path $\gamma(P)$ from P_0 to P on $\mathcal{R}_g$; then the coset $\hat{\gamma}(P) \bmod \Pi_{2g}$ in $\mathrm{Jac}(\mathcal{R}_g)$ is well determined only by P.

A meromorphic function on the Jacobian variety naturally gives a meromorphic function of g variables on $\hat{\mathrm{H}}^{(1.0)}(\mathcal{R}_g) \simeq \mathbb{C}^g$ which admits elements of Π_{2g} as its periods. It is called an Abelian function associated with the hyperelliptic differential form $dw/\sqrt{P(w)}$.

Remark: If g is greater than 1, then, in general, there might not exist g independent meromorphic functions on a complex torus $\mathbb{C}^g/\Pi$ of dimension g where Π is a $\mathbb{Z}$-module of rank $2g$ in $\mathbb{C}^g$ which generates $\mathbb{C}^g$ over $\mathbb{R}$. In case where there are g independent meromorphic functions, the complex torus is embeddable into a projective space and is an algebraic variety (by

Chow's theorem); it is called an Abelian variety. A criterion for a complex torus to be an Abelian variety will be given in §6.1 below.

6. Complex multiplication on Abelian varieties

In the last section, first we find a criterion for a complex torus to be an Abelian variety. We only present 'Riemann relations' here. For the detail, see Birkenhake and Lange [BL-2010], for example. Then we give a definition of complex multiplication on an Abelian variety which includes the one we gave in §3.2 for an elliptic curve because an elliptic curve is an Abelian variety of dimension 1.

6.1. *A criterion for a complex torus to be an Abelian variety*

Let Π be a $\mathbb{Z}$-module of rank $2g, g \in \mathbb{Z}, g \geq 1$, in $\mathbb{C}^g$ with the property $\mathbb{R} \cdot \Pi = \mathbb{C}^g$. Fix a $\mathbb{C}$-basis $\mathbf{b}_1, \ldots, \mathbf{b}_g$ of the vector space $\mathbb{C}^g$, and a set of $2g$ generators $\gamma_1, \ldots, \gamma_{2g}$ of $\mathbb{Z}$-module Π over $\mathbb{Z}$. Then a $g \times 2g$ matrix $\Omega \in \mathrm{M}_{g\times 2g}(\mathbb{C})$ is uniquely determined by the equation

$$(\gamma_1, \ldots, \gamma_{2g}) = (\mathbf{b}_1, \ldots, \mathbf{b}_g) \cdot \Omega.$$

This is called a *period matrix* for Π. If we take another set of generators of Π, then we get another period matrix of form ΩB with $B \in \mathrm{Gl}_{2g}(\mathbb{Z})$.

Definition (Riemann relations): let Ω be a period matrix and $A \in \mathrm{Gl}_{2g}(\mathbb{Z})$ be an alternating matrix: ${}^tA = -A$. Then the two relations for Ω and A,

(1) $\Omega\, A^{-1}\, {}^t\Omega = 0$,

(2) $\sqrt{-1}\Omega\, A^{-1}\, {}^t\bar{\Omega} > 0$,

are called *Riemann relations*: here the relation (2) means that the Hermitian matrix is positive definite where $\bar{\Omega}$ is the complex conjugate of Ω.

Theorem: Let the notation and the assumptions be as above. Then the complex torus $\mathbb{C}^g/\Pi$ is an *Abelian variety* if and only if a period matrix Ω for Π has an alternating matrix $A \in \mathrm{Gl}_{2g}(\mathbb{Z})$ for which the Riemann relations are satisfied.

For the proof, see [BL-2010, Ch. 8], for example.

Now let the notation and the assumptions be as in §5.2. Suppose that the closed paths $\gamma_1, \gamma_2, \ldots, \gamma_{2g-1}, \gamma_{2g}$ on the Riemann surface $\mathcal{R}_g$ are naturally chosen, for example, as in [BL-2010, Ch. 11, p.317] and put

$$c_{i,j} := \int_{\gamma_j} \frac{w^{i-1}\,dw}{\sqrt{P(w)}}, \quad i = 1, \ldots, g, \ j = 1, \ldots, 2g,$$

and $\Omega := (c_{i,j}) \in \mathrm{M}_{g,2g}(\mathbb{C})$.

Proposition: Let the notation and the assumptions be as above. Then the period matrix Ω has an alternating matrix $A \in \mathrm{Gl}_{2g}(\mathbb{Z})$ for which Riemann relations are satisfied. Hence especially, the Jacobian variety $\mathrm{Jac}(\mathcal{R}_g)$ is an Abelian variety.

For the proof, see [BL-2010, Ch. 11, §11.1], for example.

6.2. *Complex multiplication on Abelian varieties*

Suppose that the complex torus $A := \mathbb{C}^g/\Pi$ of dimension g over $\mathbb{C}$ is an Abelian variety. Then the endomorphism algebra $\mathrm{End}_{\mathbb{Q}}(A) = \mathrm{End}(A) \otimes_{\mathbb{Z}} \mathbb{Q}$ has a positive involution and is semi-simple. Here we do not explain the context clearly, and ask an interested reader to see [BL-2010, Ch. 5, §5.5], for example.

Definition 1: A *CM field* is a totally imaginary quadratic extension of a totally real number field of finite degree over $\mathbb{Q}$.

Proposition: An algebraic number field F of finite degree over $\mathbb{Q}$ has a positive involution if and only if F is either a totally real number field of finite degree or a CM field.

Definition 2 (Complex multiplication): An Abelian variety A of dimension g over $\mathbb{C}$ is said to have complex multiplication if $\mathrm{End}_{\mathbb{Q}}(A)$ is isomorphic to a CM field of degree $2g$.

Remark 1: For an arbitrary CM field F of degree $2g$ over $\mathbb{Q}$, there always exists an Abelian variety A of dimension g over $\mathbb{C}$ whose endomorphism algebra is isomorphic to the given CM field F: $\mathrm{End}_{\mathbb{Q}}(A) \simeq F$. More generally, let D be a division algebra which is central over a CM field F and has a positive involution. Then the full matrix algebra $\mathrm{M}_n(D)$ also has a positive involution. Furthermore, there always exists such an Abelian variety A as $\mathrm{End}_{\mathbb{Q}}(A)$ is isomorphic to $\mathrm{M}_n(D)$. (See Birkenhake and Lange [BL-2010, Ch. 9] and Miyake [Mi-1971]. Note that the words 'complex multiplication' in [BL-2010, Ch. 9, §9.6] is used in a wider sense than the one defined above.)

Remark 2: Shimura [Sh-1997] is a basic reference on the modern theory of Abelian varieties with complex multiplication. Arithmetic nature of Abelian varieties with complex multiplication is indispensable for arithmetic construction of moduli spaces of Abelian varieties, as is shown in [Sh-1997, §26]. In Miyake [Mi-1971], for example, a system of canonical models, that is, an 'arithmetic system' of Shimura varieties is also constructed by utilizing moduli of Abelian varieties with certain additional structures.

References

Ab-1826a. Abel, N., Démonstration de l'impossibilité de la résolution algébrique des équations générales qui passent le quatrième degré, *Jour. reine angew. Math.* **Bd.1** (1826), 65–84 (in German); *Œuvres complètes* I, 66–94 (in French).

Ab-1826b. Abel, N., Mémoire sur une propriété générale d'une classe trés étendue de fonctions transcendantes, *Œuvres complés de Niels Henrik Abel, Nouvelle Édition* I, 1881, pp.145–211.

Ab-1827. Abel, N., Recherches sur les fonctions elliptiques, *Jour. reine angew. Math.* **Bd.2** (1827), 101–181, **Bd.3** (1828), 160–190; *Œuvres complètes* I, 263–388.

Ab-1828a. Abel, N., Solution d'un problème général concernant la transformation des fonctions elliptiques, *Astronomische Nachr.* **Bd.6** (1828); *Œuvres complètes* I, 403–428.

Ab-1828b. Abel, N., Addition au mémoire précédent, *Astronomische Nachr.* **Bd.7** (1829); *Œuvres complètes* I, 429–443.

Ab-1828c. Abel, N., Remarques sur quelques propriétés générales d'un certaine sorte de fonctions transcendantes, *Jour. reine angew. Math.* **Bd.3** (1828), 313–323; *Œuvres complètes* I, 444–456.

Ab-1829a. Abel, N., Mémoire sur une classe particulière d'equations résolubles algébriquement, *Jour. reine angew. Math.* **Bd.4** (1829), 131–156; *Œuvres complètes* I, 478–507.

Ab-1829b. Abel, N., Démonstration d'une propriété d'une certaine classe de fonctions transcendantes, *Jour. reine angew. Math.* **Bd.4** (1829), 200–201; *Œuvres complètes* I, 515–517.

Ab-1881. Abel, N., *Œuvres complètes* I, II, Christiania, 1881; Reprint from Johnson Reprint, New York, 1965.

Ar-1927. Artin, E., Beweis des allgemeinen Reziprozitätsgesetz, *Abh. Math. Sem. Univ. Hamburg* **Bd.5** (1927), 353–363; *Collected Papers*, 131–141.

Ar-1965. Artin, E., *Collected Papers*, Addison Wesley, 1965.

BL-2010. Birkenhake, Christina and Herbert Lange, *Complex Abelian Varieties*, Grundlehren der mathematischen Wissenschaften, Second, Augmented Edition, Springer-Verlag, Berlin, Heidelberg, 2010.

Ga-1801. Gauss, C.F., *Disquisitiones arithmeticae*, Leibzig, 1801; Gauss Werke I, Göttingen, 1863.

Ga-1870. Gauss, C.F., *Werke* I–XII, Göttingen, 1870–1926.

Ja-1829a. Jacobi, C.G.J., *Fundamenta Nova Theoriae Functionum Ellipticarum*, Regiomonti, 1829; *Gesam. Werke* I, 49–239.

Ja-1829b. Jacobi, C.G.J., Correspondence Mathématique avec Legendre, le 14 juin, le 16 juillet et le 19 août, 1829, *Jour. reine angew. Math.* **Bd.80** (1875), 265–279; *Gesam. Werke* I, 447–452.

Ja-1835. Jacobi, C.G.J., De functionibus duarum variabilium quadrupliciter periodicis, quibus theoria transcendentium Abelianarum innititur, *Jour. reine angew. Math.* **Bd.13** (1835), 55–78; *Gesam. Werke* II, 23–50.

Ja-1846. Jacobi, C.G.J., Note sur les fonctions Abéliennes, *Bulletin de la classe physico-mathématique de l'Académie impériale des sciences de St-Pétersbourg*, Tome II, N° 7; *Jour. reine angew. Math.* **Bd.30** (1846), 183–184; *Gesam. Werke* II, 83–86.

Ja-1881. Jacobi, C.G.J., *Gesammelte Werke* I, II, III, IV, V, VI, VII, VIII, 1881–1891; Chelsea, New York, 1969.

Kr-1853. Kronecker, L., Über die algebraisch auflösbaren Gleichungen, *Monatsber. kgl. Preuss. Acad. Wiss. Berlin* (1853), 365–374; *Werke* IV, 1–11.

Kr-1857a. Kronecker, L., Über die elliptische Functionen, für welche complex Multiplication stattfindet, *Monatsber. kgl. Preuss. Acad. Wiss. Berlin* (1857), 455–460; *Werke* IV, 177–183.

Kr-1857b. Kronecker, L., Brief von L. Kronecker an G. L. Dirichlet, 17. Mai 1857, *Nachr. kgl. Akad. Wiss. Göttingen* (1885), 253–297; *Werke* V, 418–421.

Kr-1862. Kronecker, L., Über die complex Multiplication der elliptischen Functionen, *Monatsber. kgl. Preuss. Acad. Wiss. Berlin* (1862), 363–372; *Werke* IV. 207–217.

Kr-1880. Kronecker, L., Auszug aus einem Briefe von L. Kroncker an R. Dedekind, 15. März, 1880, *Werke* V, 455–457.

Kr-1895. Kronecker, L., *Mathematische Werke* I–V, Leipzig, 1895–1930; Reprint, Chelsea, New York, 1968.

La-1972. Lang, Serge, *Introduction to Algebraic and Abelian Functions*, Addison–Wesley, Reading, Massachusetts, 1972.

La-1987. Lang, Serge, *Elliptic Functions*, Graduate Texts in Mathematics, Vol. 112, Second Edition, Springer-Verlag, New York, 1987.

Le-1766. Legendre, A.-M., Mémoire sur les Intégrations par arcs d'ellipse (as le Gendre), and Second Mémoire sur les Intégrations par arcs d'ellipse.

Le-1793. Legendre, A.-M., Mémoire sur les Transcendantes elliptiques ..., lu à la ci-devant Académie des Sciences en 1792, A Paris, l'an deuxième de la République.

Le-1811. Legendre, A.-M., *Exercices du Calcul Intégral I, II, III*, 1811, 1817, 1819.

Le-1825. Legendre, A.-M., *Traité des Fonctions Elliptiques I, II, III*, 1825, 1826, 1830.

Mi-1971. Miyake, K., Models of certain automorphic function fields, *Acta Math.* **126** (1971), 245-307.

Mi-2011. Miyake, K., Takagi's Class Field Theory — From where? and to where? —, *RIMS Kôkyûroku Bessatsu,* **B25** (2011), 125–160.

Sh-1971. Shimura, G., *Introduction to the Arithmetic Theory of Automorphic Functions*, Iwanami Shoten and Princeton Univ. Press, Tokyo, 1971.

Sh-1997. Shimura, G., *Abelian Varieties with Complex Multiplication and Modular Functions*, Princeton Univ. Press, 1997.

St-1994. Stepanov, Serguei A., *Arithmetic of Algebraic Curves*, Consult. Bureau, Plenum Publ. Corporation. New York, 1994.

Ta-1920a. Takagi, T., Über eine Theorie des relativ Abel'schen Zahlkörpers, *J. Coll. Sci. Tokyo* **41** (1920), 1–133; *Collected Papers* (2nd edit.), 73–167.

Ta-1920b. Takagi, T., Sur quelques théorèmes généraux de la théorie des nombres algébriques, *Comptes rendus du congrés internat. math., Strasbourg* (1920), 185–188; *Collected Papers* (2nd edit.), 168–171.

Ta-1973. Takagi, T., *The Collected Papers*, Iwanami Shoten, 1973; the 2nd edit., Springer-Verlag, 1990.

PROBLEMS ON COMBINATORIAL PROPERTIES OF PRIMES

ZHI-WEI SUN*

Department of Mathematics, Nanjing University, Nanjing 210093, People's Republic of China
E-mail: zwsun@nju.edu.cn

1. Introduction

Prime numbers play important roles in number theory. For $x > 0$ let $\pi(x)$ denote the number of primes not exceeding x. The celebrated Prime Number Theorem states that

$$\pi(x) \sim \mathrm{Li}(x) \quad \text{as } x \to +\infty,$$

where $\mathrm{Li}(x) = \int_2^x \frac{dt}{\log t} \sim \frac{x}{\log x}$. This has the following equivalent version:

$$p_n \sim n \log n \quad \text{as } n \to +\infty,$$

where p_n denotes the n-th prime. To get sharp estimations for $\pi(x)$ is a main research topic in analytic number theory. It is known (cf. [10]) that under Riemann's Hypothesis we have

$$\pi(x) = \mathrm{Li}(x) + O(\sqrt{x}\log x) \quad \text{and} \quad p_{n+1} - p_n = O(\sqrt{p_n}\log p_n).$$

For convenience, we also set $\pi(0) = 0$.

Many number theorists generally consider primes irregular and only focus on their asymptotic behaviours. In contrast with the great achievements on the asymptotic behaviors of $\pi(x)$ and p_n (see [14] for a recent breakthrough on prime gaps), almost nobody has investigated combinatorial properties of primes seriously and systematically.

Surprisingly, we find that the functions $\pi(x)$ and p_n have many unexpected combinatorial properties depending on their exact values. Also, partition functions arising from combinatorics have nice connections with

*Supported by the National Natural Science Foundation (grant 11171140) of China.

primes. In this paper we pose 60 typical conjectures in this direction. The next section contains 25 conjectures on combinatorial properties of $\pi(x)$, while Section 3 contains 25 conjectures on combinatorial properties of the function p_n. Section 4 is devoted to 10 conjectures on primes related to partition functions. The reader may also consult [12] for the author's previous conjectures on alternating sums of consecutive primes.

The 60 selected conjectures in Sections 2–4 are somewhat incredible. Nevertheless, our numerical computations and related graphs in [11] provide strong evidences to support them. The author would like to offer 1000 Chinese dollars as the prize for the first complete solution to any one of the 60 conjectures. We hope that the problems here might interest some number theorists and stimulate further research, but the solutions to most of them might be beyond the intelligence of human beings.

2. Combinatorial properties of $\pi(x)$ and related things

Conjecture 2.1. (2014-02-09) (i) *For any integer* $n > 1$, $\pi(kn)$ *is prime for some* $k = 1, \ldots, n$. *Moreover, for every* $n = 1, 2, 3, \ldots$, *there is a positive integer* $k < 3\sqrt{n} + 3$ *with* $\pi(kn)$ *prime.*

(ii) *Let* $n_0 = 5$, $n_1 = 3$ *and* $n_{-1} = 6$. *For each* $\delta \in \{0, \pm 1\}$ *and any integer* $n > n_\delta$, *there is a positive integer* $k < n$ *such that* $k^2 + k - 1$ *and* $\pi(kn) + \delta$ *are both prime.*

Remark 2.1. (a) We also conjecture that for any integer $n > 92$ there is a prime $p \leqslant n$ with $\pi(pn)$ prime. We have verified part (i) of Conjecture 2.1 for n up to 2×10^7, and our data and graphs for the sequence $a(n) = |\{0 < k < n : \pi(kn) \text{ is prime}\}|$ $(n = 1, 2, 3, \ldots)$ (cf. [11, A237578]) strongly support its truth. It seems that $|\{1 \leqslant k \leqslant n : \pi(kn) \text{ is prime}\}| \sim \pi(n)/2$ as $n \to +\infty$. See also [11, A237615] for part (ii) of Conjecture 2.1, and note that it is not yet proven that there are infinitely many primes of the form $x^2 + x - 1$ with $x \in \mathbb{Z}$.

(b) We also conjecture the following analogue of Conjecture 2.1 (cf. [11, A238703]): For any integers $n > m > 0$ with $m \nmid n$, there is a positive integer $k < n$ with $\lfloor kn/m \rfloor$ prime. Note that $\lfloor kn/m \rfloor$ is the number of multiples of m among $1, 2, \ldots, kn$.

Conjecture 2.2. (2014-02-10) (i) *For any positive integer* n, *there is a positive integer* $k < p_n$ *such that* $\pi(kn) \equiv 0 \pmod{n}$.

(ii) *For each positive integer* n, *the set* $\{\pi(kn) : k = 1, \ldots, 2p_n\}$ *contains a complete system of residues modulo* n.

Remark 2.2. See [11, A237597] and [11, A237643] for related sequences concerning this conjecture.

Conjecture 2.3. (2014-02-20) *Let $n > 1$ be an integer. Then $\pi(jn) \mid \pi(kn)$ for some $1 \leqslant j < k \leqslant n$ with $k \equiv 1 \pmod{j}$.*

Remark 2.3. For example, $\pi(3 \times 50) = 35$ divides $\pi(7 \times 50) = 70$ with $7 \equiv 1 \pmod 3$. We have verified the conjecture for all $n = 2, 3, \ldots, 30000$. See [11, A238224] for a related sequence.

Conjecture 2.4. (i) (2014-02-10) *For any positive integer n, there is a positive integer $k < p_n$ such that $\pi(kn)$ is a square.*

(ii) (2014-02-14) *Let n be any positive integer. Then, for some $k = 1, \ldots, n$, the number of twin prime pairs not exceeding kn is a square.*

Remark 2.4. See [11, A237598, A237612, A237840, A237879 and A237975] for some sequences related to this conjecture. Similar to part (i), we conjecture that for any integer $n > 9$ there is a positive integer $k < p_n/2$ such that $\pi(kn)$ is a triangular number. We have verified part (ii) of the conjecture for all $n = 1, \ldots, 22000$; for example, for $n = 19939$ we may take $k = 12660$ since there are exactly $1000^2 = 10^6$ twin prime pairs not exceeding $12660 \times 19939 = 252427740$.

Conjecture 2.5. (i) (2014-02-24) *For any integer $n > 5$, there is a positive integer $k < n$ with $kn + \pi(kn)$ prime.*

(ii) (2014-03-06) *If n is a positive integer, then $p_{kn} - \pi(kn)$ is prime for some $k = 1, \ldots, n$.*

Remark 2.5. See [11, A237712 and A238890] for related sequences. Part (ii) of the conjecture implies that there are infinitely many primes p with $p - \pi(\pi(p))$ prime.

Conjecture 2.6. (2014-03-07) (i) *For any integer $n > 2$, there is a prime $p \leqslant n$ with $\pi(\pi((p-1)n))$ prime.*

(ii) *Let n be any positive integer. Then $\pi(\pi(kn))$ is a square for some $k = 1, \ldots, n$. Also, there exists a positive integer $k \leqslant (n+1)/2$ such that $\pi(\pi(kn))$ is a triangular number.*

Remark 2.6. See [11, A238504, A238902 and A239884] for related data and graphs. We have verified the two assertions in Conjecture 2.6(ii) for n

up to 2×10^5 and 10^5 respectively; for example,

$$\pi(\pi(8514 \times 9143)) = \pi(4550901) = 565^2,$$
$$\pi(\pi(37308 \times 98213)) = \pi(174740922) = 3123^2,$$
$$\pi(\pi(83187 \times 192969)) = \pi(715034817) = 6082^2.$$

We guess that there are positive constants c_1 and c_2 such that

$$c_1 \leqslant \frac{|\{1 \leqslant k \leqslant n : \ \pi(\pi(kn)) \text{ is a square}\}|}{\log n} \leqslant c_2 \qquad \text{for all } n = 2, 3, \ldots.$$

Conjecture 2.7. (i) (2014-02-17) *For any integer $n > 4$ and $k = 1, \ldots, n$, we have $\pi(kn)^{1/k} > \pi((k+1)n)^{1/(k+1)}$.*

(ii) (2014-02-22) *Let n be any positive integer. Then, $\pi((k+1)n) - \pi(kn)$ (the number of primes in the interval $(kn, (k+1)n]$) is a square for some $k = 0, \ldots, n-1$.*

Remark 2.7. For any integer $n > 1$, Bertrand's postulate (first proved by Chebyshev in 1850) indicates that $\pi(2n) > \pi(n)$, and Oppermann's conjecture states that $\pi((n-1)n) < \pi(n^2) < \pi(n(n+1))$. Our computation suggests that $\pi(kn) < \pi((k+1)n)$ for any integers $n \geqslant k > 0$. A conjecture of Firoozbakht (cf. [9, p. 185]) asserts that the sequence $p_n^{1/n}$ $(n = 1, 2, 3, \ldots)$ is strictly decreasing, and the author's recent paper [13] contains many similar conjectures on monotonicity of arithmetical sequences. See also [11, A238277] for a sequence related to Conjecture 2.7(ii).

Conjecture 2.8. (i) (2014-03-16) *Let $n > 3$ be an integer. Then $\pi(pn) - \pi((p-1)n)$ is prime for some prime $p < n$. Also, there is an odd prime $p \leqslant n$ with $\pi(\frac{p+1}{2}n) - \pi(\frac{p-1}{2}n)$ prime.*

(ii) (2014-02-22) *For any integer $n > 3$, there is a number $k \in \{1, \ldots, n-1\}$ with $\pi(kn) - \pi((k-1)n)$ and $\pi((k+1)n) - \pi(kn)$ both prime.*

Remark 2.8. See [11, A239328, A239330 and A238278] for related data and graphs.

Conjecture 2.9. (2014-02-22) (i) *For any integer $n > 1$, there is a positive integer $k < n$ such that the intervals $(kn, (k+1)n)$ and $((k+1)n, (k+2)n)$ contain the same number of primes, i.e.,*

$$\pi(kn), \ \pi((k+1)n), \ \pi((k+2)n)$$

form a three-term arithmetic progression.

(ii) *For any integer $n > 4$, there is a positive integer $k < p_n$ such that*

$$\pi(kn), \ \pi((k+1)n), \ \pi((k+2)n), \ \pi((k+3)n)$$

form a four-term arithmetic progression.

Remark 2.9. See [11, A238281] for a sequence related to part (i) of this conjecture.

Conjecture 2.10. (2014-02-23) *For any positive integer n, we have*

$$|\{\pi((k+1)n)-\pi(kn):\ k=0,\ldots,n-1\}|\geqslant\sqrt{n-1},$$

and equality holds only when n is 2 or 26.

Remark 2.10. See [11, A230022] for related data and graphs.

Conjecture 2.11. (2014-02-24) *Let $n>1$ be an integer. Then, for some prime $p\leqslant p_n$, the three numbers $\pi(p),\pi(p+n),\pi(p+2n)$ form a nontrivial arithmetic progression, i.e., $\pi(p+2n)-\pi(p+n)=\pi(p+n)-\pi(p)>0$.*

Remark 2.11. See [11, A210210] for a related sequence.

Conjecture 2.12. (i) (2014-02-08) *Each integer $n>10$ can be written as $k+m$ with k and m positive integers such that $\pi(km)$ (or $\pi(k^2m)$) is prime.*

(ii) (2014-03-20) *For any integer $n>9$, there are positive integers k and m with $k+m=n$ such that $\pi(2k)-\pi(k)$ and $\pi(2m)-\pi(m)$ are both prime. Also, any integer $n>4$ can be written as a sum of two positive integers k and m such that $\pi(2k)-\pi(k)$ is a prime and $\pi(2m)-\pi(m)$ is a square.*

Remark 2.12. See [11, A237497, A239428 and A239430] for related data and graphs. As $\pi(2k+2)-\pi(k+1)-(\pi(2k)-\pi(k))\in\{0,\pm1\}$ and $\pi(2n)-\pi(n)\sim n/\log n$, there are infinitely many positive integers k with $\pi(2k)-\pi(k)$ prime. Similar to part (i), we also conjecture (cf. [11, A237531]) that for any integer $n>5$ there is a positive integer $k<n/2$ such that $\varphi(k(n-k))-1$ and $\varphi(k(n-k))+1$ are twin prime, where φ denotes Euler's totient function.

For $x\geqslant 0$, we use $\pi_2(x)$ to denote the number of twin prime pairs not exceeding x, i.e., $\pi_2(x)=|\{p\leqslant x:\ p \text{ and } p-2 \text{ are both prime}\}|$.

Conjecture 2.13. (2014-02-15) (i) *Each integer $n>5$ can be written as $k+m$ with k and m positive integers such that $\pi_2(km)$ is prime.*

(ii) *Any integer $n>8$ can be written as $k+m$ with k and m positive integers such that $\pi_2(km)-1$ and $\pi_2(km)+1$ are twin prime.*

Remark 2.13. This is an analogue of Conjecture 2.12(i) for twin prime pairs.

Conjecture 2.14. (i) (2014-02-11) *For any positive integer n, the set $\{\pi(k^2) : k = 1, \ldots, 2p_{n+1} - 3\}$ contains a complete system of residues modulo n.*

(ii) (2014-02-17) *The sequence $\sqrt[n]{\pi(n^2)}$ $(n = 3, 4, \ldots)$ is strictly decreasing.*

(iii) (2014-02-17) *For any integer $n > 0$, the interval $[\pi(n^2), \pi((n+1)^2)]$ contains at least one prime except for $n = 25, 35, 44, 46, 105$.*

Remark 2.14. Legendre's conjecture asserts that for each positive integer n there is a prime between n^2 and $(n+1)^2$.

Conjecture 2.15. (i) (2014-02-11) *For any integer $n > 8$, $\pi(k)$ and $\pi(k^2)$ are both prime for some integer $k \in (n, 2n)$.*

(ii) (2014-02-11) *There are infinitely many primes p with $\pi(p)$, $\pi(\pi(p))$ and $\pi(p^2)$ all prime.*

(iii) (2014-04-09) *For any positive integer n, there are infinitely many primes p with $\pi(kp)$ prime for all $k = 1, \ldots, n$.*

Remark 2.15. See [11, A237657, A237687 and A240604] for related sequences and data.

Conjecture 2.16. (i) (2014-02-09) *For any integer $n > 1$, $\pi(n + k^2)$ is prime for some $k = 1, \ldots, n-1$. In general, for each $a = 2, 3, \ldots$, if an integer n is sufficiently large, then $\pi(n + k^a)$ is prime for some $k = 1, \ldots, n-1$.*

(ii) (2014-02-10) *Let $n > 4$ be an integer. Then $n + \pi(k^2)$ is prime for some $k = 1, \ldots, n$.*

(iii) (2014-03-01) *If a positive integer n is not a divisor of 12, then $n^2 + \pi(k^2)$ is prime for some $1 < k < n$. For any integer $n > 4$, $\pi(n^2) + \pi(k^2)$ is prime for some $1 < k < n$. Also, for each $n = 2, 3, \ldots$ there is a positive integer $k < n$ such that $\pi((k+1)^2) - \pi(k^2)$ and $\pi(n^2) - \pi(k^2)$ are both prime.*

Remark 2.16. See [11, A237582, A237595 and A238570] for related sequences. In 2012 the author conjectured that if n is a positive integer then $n + k$ and $n + k^2$ are both prime for some $k = 0, \ldots, n$ (cf. [11, A185636]).

Conjecture 2.17. (i) (2014-02-08) *For any integer $n > 4$, there is a prime $p < n$ with $pn + \pi(p)$ prime. Moreover, for every positive integer n, there is a prime $p < \sqrt{2n}\log(5n)$ with $pn + \pi(p)$ prime.*

(ii) (2014-03-02) *For any integer $n > 2$, there is a prime $p \leqslant n$ with $2\pi(p) - (-1)^n$ and $pn + ((-1)^n - 3)/2$ both prime.*

Remark 2.17. See [11, A237453 and A238643] for related data and graphs. We have verified parts (i) and (ii) for n up to 10^8. As a supplement to part (i), we also conjecture that for every $n = 1, 2, 3, \ldots$ there is a positive integer $k < 3\sqrt{n}$ such that $kn + p_k = \pi(p_k)n + p_k$ is prime. Part (ii) implies that for any odd prime p there is a prime $q \leqslant p$ with $pq - 2$ prime. By Chen's work [2], there are infinitely many primes p with $p+2$ a product of at most two primes.

Conjecture 2.18. (2013-11-24) (i) *Every $n = 4, 5, \ldots$ can be written as $p + q - \pi(q)$, where p and q are odd primes not exceeding n.*

(ii) *For any integer $n > 7$, there is a prime $p < n$ with $n + p - \pi(p)$ also prime.*

Remark 2.18. We have verified part (i) for all $n = 4, 5, \ldots, 10^8$; for example, $9 = 7 + 5 - \pi(5)$ with 7 and 5 prime. See [11, A232463 and A232443] for related sequences.

Conjecture 2.19. (2014-02-06) (i) *For any integer $n > 2$, there is a prime $p < 2n$ with $\pi(p)$ and $2n - p$ both prime.*

(ii) *For any integer $n > 36$, we can write $2n - 1 = a + b + c$ with a, b, c in the set $\{p : p \text{ and } \pi(p) \text{ are both prime}\}$.*

Remark 2.19. Part (i) is a refinement of Goldbach's conjecture, and part (ii) is stronger than the weak Goldbach conjecture finally proved by Helfgott [7]. See [11, A237284 and A237291] for related representation functions.

Recall that a prime p with $2p + 1$ also prime is called a Sophie Germain prime.

Conjecture 2.20. (2014-02-13) (i) *For any integer $n > 4$, there is a prime $p < n$ such that $\pi(n - p)$ is a Sophie Germain prime. Also, for any integer $n > 8$ there is a prime $p < n$ such that $\pi(n - p) - 1$ and $\pi(n - p) + 1$ are twin prime.*

(ii) *For any integer $n > 4$, there is a prime $p < n$ such that $3m \pm 1$ and $3m + 5$ are all prime with $m = \pi(n - p)$. Also, for any integer $n > 8$, there is a prime $p < n$ such that $3m \pm 1$ and $3m - 5$ are all prime with $m = \pi(n - p)$.*

Remark 2.20. See [11, A237768 and A237769] for related sequences. We have verified part (i) for n up to 2×10^7.

Conjecture 2.21. (2014-02-13) (i) *For any integer $n > 4$, there is a prime $p < n$ such that the number of Sophie Germain primes among $1, \ldots, n-p$ is a Sophie Germain prime.*

(ii) *For any integer $n > 12$, there is a prime $p < n$ such that*

$$r = |\{q \leqslant n-p: \ q \text{ and } q+2 \text{ are twin prime}\}|$$

and $r+2$ are twin prime.

Remark 2.21. See [11, A237815 and A237817] for related sequences.

Conjecture 2.22. (2014-02-12) (i) *For any integer $n > 2$, there is a prime $p < n$ such that $\pi(n-p)$ is a square. Also, for any integer $n > 2$ there is a prime $p < n$ such that $\pi(n-p)$ is a triangular number.*

(ii) *For any integer $n > 2$, there is a prime $p \leqslant p_n$ such that $\pi(n+p)$ is a square.*

Remark 2.22. See [11, A237706 and A237710] for related sequences. We have verified the first assertion in part (i) for n up to 5×10^8, and guessed that the number of primes $p < n$ with $\pi(n-p)$ a square is asymptotically equivalent to $c\sqrt{n}$ with c a constant in the interval $(0.2, 0.22)$. We also conjecture the following analogue (cf. [11, A238732]) of Conjecture 2.22(i): For any integers $m > 2$ and $n > 2$, there is a prime $p < n$ such that $\lfloor (n-p)/m \rfloor$ is a square.

Conjecture 2.23. (2014-02-13) (i) *For any integer $n > 11$, there is a prime $p < n$ such that the number of Sophie Germain primes among $1, \ldots, n-p$ is a square.*

(ii) *For any integer $n \geqslant 54$, there is a prime $p < n$ such that the number of Sophie Germain primes among $1, \ldots, n-p$ is a cube.*

Remark 2.23. Part (i) is an analogue of Conjecture 2.22(i) for Sophie Germain primes. See [11, A237837] for a sequence related to part (ii).

Conjecture 2.24. (2014-03-02) (i) *For every $n = 2, 3, \ldots$, there is an odd prime $p < 2n$ such that the number of squarefree integers among $1, \ldots, \frac{p-1}{2}n$ is prime.*

(ii) *For any integer $n > 3$, there is a prime $p < n$ such that the number of squarefree numbers among $1, \ldots, n-p$ is prime.*

Remark 2.24. See [11, A238645 and A238646] for related sequences.

Conjecture 2.25. (2014-02-22) *For any integer $n > 4$, there is a number $k \in \{1, \dots, n\}$ such that the number of prime ideals of the Gaussian ring $\mathbb{Z}[i]$ with norm not exceeding kn is a prime congruent to 1 modulo 4.*

Remark 2.25. $\mathbb{Z}[i]$ is a principal ideal domain, and any prime ideal P of it has the form (p) with p a rational prime congruent to 3 modulo 4 or $p = a + bi$ with $N(p) = a^2 + b^2$ a rational prime not congruent to 3 modulo 4. (Cf. [8, p. 120].) So, the number of prime ideals of $\mathbb{Z}[i]$ with norm not exceeding x actually equals

$$\pi(\sqrt{x}) + |\{\sqrt{x} < p \leqslant x : \ p \text{ is a prime with } p \not\equiv 3 \pmod 4\}|.$$

3. Combinatorial properties involving the function p_n

Conjecture 3.1. (Unification of Goldbach's Conjecture and the Twin Prime Conjecture, 2014-01-29) *For any integer $n > 2$, there is a prime q with $2n - q$ and $p_{q+2} + 2$ both prime.*

Remark 3.1. We have verified this for n up to 2×10^8. See [11, A236566] for a related sequence. Note that the conjecture implies the Twin Prime Conjecture. In fact, if all primes q with $p_{q+2} + 2$ prime are smaller than an even number $N > 2$, then for any such a prime q the number $N! - q$ is composite since $N! - q \equiv 0 \pmod q$ and $N! - q \geqslant q(q+1) - q > q$.

Conjecture 3.2. (Super Twin Prime Conjecture, 2014-02-05) *Any integer $n > 2$ can be written as $k+m$ with k and m positive integers such that p_k+2 and $p_{p_m} + 2$ are both prime.*

Remark 3.2. We have verified the conjecture for n up to 10^9. See [11, A218829, A237259, A237260] for related sequences. If $p, p+2$ and $\pi(p)$ are all prime, then we call $\{p, p+2\}$ a *super twin prime* pair. Conjecture 3.2 implies that there are infinitely many super twin prime pairs. In fact, if all those positive integers m with $p_{p_m} + 2$ prime are smaller than an integer $N > 2$, then by Conjecture 3.2, for each $j = 1, 2, 3, \dots$, there are positive integers $k(j)$ and $m(j)$ with $k(j) + m(j) = jN$ such that $p_{k(j)} + 2$ and $p_{p_{m(j)}} + 2$ are both prime, and hence $k(j) \in ((j-1)N, jN)$ since $m(j) < N$; thus

$$\sum_{j=1}^{\infty} \frac{1}{p_{k(j)}} \geqslant \sum_{j=1}^{\infty} \frac{1}{p_{jN}},$$

which is impossible since the series on the right-hand side diverges while the series on the left-hand side converges by Brun's theorem on twin primes (cf. [3, p. 14]).

Conjecture 3.3. (2014-01-28) *Any integer $n > 2$ can be written as $k+m$ with k and m positive integers such that both $\{6k \pm 1\}$ and $\{p_m, p_m + 2\}$ are twin prime pairs.*

Remark 3.3. Clearly this implies the Twin Prime Conjecture. We have verified Conjecture 3.3 for n up to 2×10^7. See [11, A236531] for related data and graphs.

Conjecture 3.4. (2014-02-07) *Any integer $n > 1$ can be written as $k+m$ with k and m positive integers such that $p_k^2 - 2$, $p_m^2 - 2$ and $p_{p_m}^2 - 2$ are all prime.*

Remark 3.4. We have verified this for n up to 10^8. See [11, A237413 and A237414] for related sequences. It is not yet proven that there are infinitely many primes of the form $x^2 - 2$ with $x \in \mathbb{Z}$.

Conjecture 3.5. (i) (2014-11-25) *The set*

$$\{k+m: \ 0 < k < m < n, \text{ and } p_k, p_m, p_n \text{ form an arithmetic progression}\}$$

coincides with $\{5, 6, 7, \ldots\}$.

(ii) (2014-02-22) *For every $n = 1, 2, 3, \ldots$, there is a positive integer $k \leqslant 3p_n + 8$ such that $p_{kn}, p_{(k+1)n}, p_{(k+2)n}$ form a three-term arithmetic progression.*

Remark 3.5. Recall that the Green-Tao theorem (cf. [4]) asserts that there are arbitrarily long arithmetic progressions of primes. See [11, A232502 and A238289] for related sequences.

Conjecture 3.6. (2014-03-01) (i) *For any integer $n > 6$, there is a number $k \in \{1, \ldots, n\}$ with $p_{kn} + 2$ (or $p_{k^2 n} + 2$) prime. Moreover, for every $n = 1, 2, 3, \ldots$ there is a positive integer $k < 3\sqrt{n} + 6$ with $p_{kn} + 2$ prime.*

(ii) *For any positive integer n, there is a number $k \in \{1, \ldots, n\}$ such that $2k + 1$ and $p_{kn}^2 - 2$ are both prime.*

Remark 3.6. Clearly part (i) is stronger than the Twin Prime Conjecture, while part (ii) implies that there are infinitely many primes p with $p^2 - 2$ prime. See [11, A238573 and A238576] for related data and graphs.

Conjecture 3.7. (i) (2013-12-01) *There are infinitely many positive integers n such that*

$$n \pm 1, \ p_n \pm n, \ np_n \pm 1$$

are all prime.

(ii) (2014-01-20) *There are infinitely many primes q with $p_q^2 + 4q^2$ and $q^2 + 4p_q^2$ both prime.*

Remark 3.7. For part (i), the first such a number n is 22110; see [11, A232861] for a list of the first 2000 such numbers n. See also [11, A236193] for a list of the first 10000 suitable primes q in part (ii) of Conjecture 3.7.

Conjecture 3.8. (i) (2013-12-07) *For every $n = 2, 3, \ldots$, there is a positive integer $k < n$ with $kp_{n-k} + 1$ prime. Also, for any integer $n > 2$, there is a positive integer $k < n$ with $kp_{n-k} - 1$ prime.*

(ii) (2013-12-11) *Let $n > 5$ be an integer. Then $p_k p_{n-k} - 6$ is prime for some $0 < k < n$.*

Remark 3.8. See [11, A233296 and A233529] for related data and graphs. By Chen's work [2], there are infinitely many primes p with $p+6$ a product of at most two primes.

Conjecture 3.9. (2013-11-23) *Let $n > 6$ be an integer. Then $p_k + p_{n-k} - 1$ is prime for some $k = 1, \ldots, n-1$. Also, $p_k^2 + p_{n-k}^2 - 1$ is prime for some $k = 1, \ldots, n-1$.*

Remark 3.9. See [11, A232465] for a related sequence.

Conjecture 3.10. (2013-12-10) (i) *For any integer $n > 3$, there is a positive integer $k < n$ such that $p_k^2 + 4p_{n-k}^2$ is prime.*

(ii) *Let $n > 10$ be an integer. Then there is a positive integer $k < n$ with $p_k^3 + 2p_{n-k}^3$ prime. Also, $p_k^3 + 2p_{n-k}^2$ is prime for some $0 < k < n$.*

Remark 3.10. See [11, A233439] for a sequence related to part (i). In 2001 Heath-Brown [6] proved that there are infinitely many primes of the form $x^3 + 2y^3$ with $x, y \in \mathbb{Z}$.

Conjecture 3.11. (2014-03-01) (i) *If a positive integer n is not a divisor of 6, then $p_q^2 + (p_n - 1)^2$ is prime for some prime $q < n$. Also, for any positive integer $n \neq 1, 2, 9$, there is a prime $q < n$ with $(p_q - 1)^2 + p_n^2$ prime.*

(ii) *For every $n = 2, 3, \ldots$, there is a positive integer $k < n$ with $p_n^3 + 2p_k^3$ prime.*

Remark 3.11. See [11, A238585] for a sequence related to the first assertion in the conjecture.

Conjecture 3.12. (i) (2013-12-05) *Any integer $n > 7$ can be written as $k + m$ with k and m positive integers such that $2^k + p_m$ is prime.*

(ii) (2013-12-06) *Any integer $n > 3$ can be written as $k + m$ with k and m positive integers such that $k! + p_m$ is prime.*

Remark 3.12. See [11, A233150 and A233206] for related sequences. We have verified parts (i) and (ii) for n up to 3×10^7 and 10^7 respectively. For example, for $n = 28117716$ we may take $k = 81539$ and $m = 28036177$ so that $2^k + p_m$ is prime, also $11 = 4 + 7$ with $4! + p_7 = 24 + 17 = 41$ prime. Part (i) was motivated by the author's conjecture (cf. [11, A231201]) that any integer $n > 1$ can be written as a sum of two positive integers k and m with $2^k + m$ prime. We also conjecture that for any integer $n > 1$ there is a number $k \in \{1, \ldots, n\}$ with $k!n - 1$ (or $k!n + 1$) prime.

Conjecture 3.13. (i) (2013-12-05) *Any integer $n > 2$ can be written as $k + m$ with $m > k > 0$ integers such that $\binom{2k}{k} + p_m$ is prime.*

(ii) (2014-03-19) *For any integer $n > 4$, there exists an integer $1 < k < \sqrt{n} \log n$ such that $p_n + \binom{p_k - 1}{(p_k - 1)/2}$ is prime.*

Remark 3.13. We have verified parts (i) and (ii) for n up to 10^8 and 10^7 respectively. See [11, A233183 and A239451] for related data and graphs.

Conjecture 3.14. (2014-01-21) *For any integer $n \geqslant 20$, there is a positive integer $k < n$ such that $m = \varphi(k) + \varphi(n - k)/8$ is an integer with $\binom{2m}{m} + p_m$ prime.*

Remark 3.14. This implies that there are infinitely many positive integers m with $\binom{2m}{m} + p_m$ prime. (By Stirling's formula, $\binom{2m}{m} \sim 4^m/\sqrt{m\pi}$ as $m \to +\infty$.) See [11, A236241] for the corresponding representation function, and [11, A236242] for a list of 52 values of m with $\binom{2m}{m} + p_m$ prime. For example, when $m = 30734$ the number $\binom{2m}{m} + p_m$ is a prime with 18502 decimal digits.

Conjecture 3.15. (i) (2013-12-29) *Any integer $n > 9$ can be written as $k + m$ with k and m positive integers such that $q = k + p_m$ and $p_q - q + 1$ are both prime.*

(ii) (2014-03-05) *Each integer $n > 1$ can be written as $k + m$ with k and m positive integers such that*

$$p_{p_k} - p_k + 1, \quad p_{p_{2k+1}} - p_{2k+1} + 1 \quad \textit{and} \quad p_{p_m} - p_m + 1$$

are all prime.

(iii) (2014-03-06) *For any positive integer n, there is a number $k \in \{1, \ldots, n\}$ such that $p_{p_k} - p_k + 1$ and $p_{p_{kn}} - p_{kn} + 1$ are both prime.*

(iv) (2014-03-06) *There are infinitely many primes* q *with* $p_q - q + 1$ *and* $p_{q'} - q' + 1$ *both prime, where* q' *is the first prime after* q.

Remark 3.15. See [11, A234694, A238766, A238878] for data and graphs related to parts (i)-(iii). We have verified part (ii) of the conjecture for all $n = 2, 3, \ldots, 10^7$. See also [11, A234695] for the first 10000 primes q with $p_q - q + 1$ also prime, and [11, A238814] for the first 10000 primes q with $p_q - q + 1$ and $p_{q'} - q' + 1$ both prime.

Conjecture 3.16. (2014-03-03) *Let* $m > 0$ *and* $n > 2m + 1$ *be integers. If* $m = 1$ *and* $2 \mid n$, *or* $m = 3$ *and* $n \not\equiv 1 \pmod 6$, *or* $m \in \{2, 4, 5, \ldots\}$, *then there is a prime* $p < n$ *such that* $q = \lfloor (n-p)/m \rfloor$ *and* $p_q - q + 1$ (*or* $q^2 - 2$) *are both prime.*

Remark 3.16. In the case $m = 1$, this is a refinement of Goldbach's conjecture. When $m = 2$, it is stronger than Lemoine's conjecture which states that any odd number $n > 5$ can be written as $p + 2q$ with p and q both prime. Conjecture 3.16 in the case $m > 2$ is completely new. We have verified the conjecture for all $m = 1, \ldots, 40$ and $n = 2m + 2, \ldots, 10^6$. See [11, A235189, A238134 and A238701] for related data and graphs.

Conjecture 3.17. (2014-01-30) *Any odd number greater than* 5 *can be written as a sum of three elements of the set*

$$\{q : \text{ both } q \text{ and } p_q - q + 1 \text{ are prime}\}.$$

Remark 3.17. This is stronger than the weak Goldbach conjecture finally proved by Helfgott [7]. See [11, A236832] for the corresponding representation function.

Conjecture 3.18. (2014-01-19) *For any integer* $n \geqslant 32$, *there is a positive integer* $k < n - 2$ *such that* $q = \varphi(k) + \varphi(n-k)/2 + 1$ *and* $p_q - q \pm 1$ *are all prime.*

Remark 3.18. See [11, A236097 and A236119] for related sequences. The conjecture implies that there are infinitely twin prime pairs of the form $p_q - q \pm 1$ with q prime.

Conjecture 3.19. (2014-01-17) *For any integer* $n \geqslant 38$, *there is a positive integer* $k < n$ *such that* $q = \varphi(k) + \varphi(n-k)/3 + 1$, $r = p_q - q + 1$ *and* $s = p_r - r + 1$ *are all prime.*

Remark 3.19. See [11, A235924 and A235925] for related sequences. The conjecture implies that there are infinitely primes q with $r = p_q - q + 1$ and $s = p_r - r + 1$ both prime.

Conjecture 3.20. (2014-01-17) *For each $m = 2, 3, \ldots$, there is a prime chain $q_1 < \ldots < q_m$ of length m such that $q_{k+1} = p_{q_k} - q_k + 1$ for all $0 < k < m$.*

Remark 3.20. For such chains of length $m = 4, 5, 6$, see [11, A235934, A235935 and A235984]. We also have some other conjectures similar to Conjecture 3.20, see, e.g., [11, A236066 and A236481].

Conjecture 3.21. (i) (2014-03-06) *For any integer $n > 5$, there is a positive integer $k < n$ such that $2k + 1$ and $p_{kn} + kn$ are both prime.*

(ii) (2014-01-04) *Any integer $n > 8$ can be written as $k(k + 1)/2 + m$ with k and m positive integers such that $p_{k(k+1)/2} + \varphi(m)$ is prime.*

Remark 3.21. See [11, A238881 and A235061] for related sequences. Motivated by part (i), we conjecture that for any integer-valued polynomial $P(x)$ with positive leading coefficient there are infinitely many positive integers n with $p_n + 2P(n)$ prime. When $P(x)$ is constant, this reduces to a conjecture of de Polignac.

Conjecture 3.22. (2013-12-16) (i) *Any integer $n > 100$ can be written as $k^2 + m$ with k and m positive integers such that $\varphi(k^2) + p_m$ is prime.*

(ii) *If an integer $n > 6$ is not equal to 18, then it can be written as $k^2 + m$ with k and m positive integers such that $\sigma(k^2) + p_m - 1$ is prime, where $\sigma(j)$ is the sum of all positive divisors of j.*

Remark 3.22. See [11, A236548] for a sequence related to part (i). Conjecture 3.22 was motivated by the author's conjecture (cf. [11, A233544]) that any integer $n > 1$ can be written as $k^2 + m$ with $\sigma(k^2) + \varphi(m)$ prime, where k and m are positive integers with $m \geqslant k^2$.

Conjecture 3.23. (2014-02-01) (i) *For any integer $n > 13$, there is a prime $q < n$ such that $q + 2$ and $p_{n-q} + q + 1$ are both prime.*

(ii) *If a positive integer n is not a divisor of 12, then there is a prime $q < n$ such that $3(p_{n-q} + q) - 1$ and $3(p_{n-q} + q) + 1$ are twin prime.*

Remark 3.23. See [11, A236831 and A182662] for related sequences.

Conjecture 3.24. (2014-05-22) (i) *Any integer $n > 3$ can be written as $a + b$ with a and b in the set*

$$\{k > 0 : \text{ the inverse of } k \text{ mod } p_k \text{ is prime}\},$$

where the inverse of k mod p_k refers to the unique $x \in \{1, \ldots, p_k - 1\}$ with $kx \equiv 1 \pmod{p_k}$.

(ii) *Every $n = 2, 3, 4, \ldots$ can be written as $a + b$ with a and b in the set*

$$\{k > 0 : \ k \text{ is a primitive root modulo } p_k\}.$$

Remark 3.24. See [11, A242753 and A242748] for related sequences. We have verified parts (i) and (ii) for n up to 10^8 and 3×10^5 respectively. We also conjecture that for any prime $p > 5$ there is a positive square $k^2 < p$ such that the inverse of k^2 mod p is prime (cf. [11, A242425]), and that any integer $n > 7$ can be written as $k + m$ with $k, m \in \{2, 3, \ldots\}$ such that the least positive residue of p_k modulo k is prime and the least positive residue of p_m modulo m is a square (cf. [11, A242950]).

Conjecture 3.25. (2014-06-01) *Let $n > 6$ be an integer. Then there is a prime $p < n$ such that pn is a primitive root modulo p_n. Also, there is a prime $q < n$ such that $q(n - q)$ is a primitive root modulo p_n.*

Remark 3.25. See [11, A243164 and A243403] for related data and graphs. We have verified Conjecture 3.25 for all $n = 7, \ldots, 2 \times 10^5$.

4. On primes related to partition functions

For $n = 1, 2, 3, \ldots$, let $p(n)$ denote the number of ways to write n as a sum of positive integers with the order of addends ignored. The function $p(n)$ is called the *partition function.* For each positive integer n, let $q(n)$ denote the number of ways to write n as a sum of *distinct* positive integers with the order of addends ignored. The function $q(n)$ is usually called the *strict partition function.* It is known that

$$p(n) \sim \frac{e^{\pi\sqrt{2n/3}}}{4\sqrt{3}n} \quad \text{and} \quad q(n) \sim \frac{e^{\pi\sqrt{n/3}}}{4(3n^3)^{1/4}} \qquad \text{as } n \to +\infty$$

(cf. [5] and [1, p. 826]). So both $p(n)$ and $q(n)$ grow eventually faster than any polynomial in n.

Conjecture 4.1. (i) (2014-02-27) *Let n be any positive integer. Then $p(n) + k$ is prime for some $k = 1, \ldots, n$. Also, $q(n) + k$ is prime for some $k = 1, \ldots, n$.*

(ii) (2014-02-28) *Let $n > 1$ be an integer. Then $p(n) + p(k) - 1$ is prime for some $0 < k < n$, and $p(k) + q(n)$ is prime for some $0 < k < n$. Also, for any integer $n > 7$, there is a positive integer $k < n$ with $n + p(k)$ prime.*

(iii) (2014-03-12) *Let $n > 1$ be an integer. Then there exists a number $k \in \{1, \ldots, n-1\}$ such that $kp(n)(p(n)-1)+1$ is prime. Also, we may replace $kp(n)(p(n)-1)+1$ by $p(k)p(n)(p(n)-1)+1$ or $p(k)p(n)(p(n)+1)-1$.*

Remark 4.1. See [11, A238457, A238509, A239209 and A239214] for related data and graphs. For part (i) or part (ii), we have verified the first assertion for n up to 1.5×10^5. For part (iii), we have verified the first assertion for n up to 10^5. Conjecture 4.1 might be helpful in finding large primes.

Conjecture 4.2. (2014-02-27) (i) *For any integer $n > 2$, there is a prime $q < n$ with $2p(n-q)+1$ prime. Also, for every $n = 4, 5, \ldots$, there is a prime $q < n$ with $2p(n-q)-1$ prime.*

(ii) *For each integer $n > 2$, there is a prime $p < n$ with $q(n-p)+1$ prime. Also, for any integer $n > 6$, there is a prime $p < n$ with $q(n-p)-1$ prime.*

Remark 4.2. This is an analogue of Conjecture 2.20. We have verified the conjecture for n up to 10^5. See [11, A238458 and A238459] for related sequences.

Conjecture 4.3. (i) (2013-12-26) *For any integer $n > 127$, there is a positive integer $k < n-2$ such that $p(k + \varphi(n-k)/2)$ is prime.*

(ii) (2013-12-28) *For any integer $n > 727$, there is a positive integer $k < n-2$ such that $q = \varphi(k) + \varphi(n-k)/2 + 1$ and $p(q-1)$ are both prime.*

Remark 4.3. Clearly, part (ii) implies that there are infinitely many primes q with $p(q-1)$ prime. We have verified parts (i) and (ii) for n up to 25000 and 56000 respectively. See [11, A234470, A234567 and A234569] for related data and graphs.

Conjecture 4.4. (i) (2014-03-13) *For each integer $n > 3$, there is a number $k \in \{1, \ldots, n\}$ with $p(n+k)+1$ prime. Also, for any integer $n > 15$, there is a number $k \in \{1, \ldots, n\}$ with $p(n+k)-1$ prime.*

(ii) (2013-12-26) *Any integer $n > 5$ can be written as $k+m$ with $k, m \in \{3, 4, \ldots\}$ such that $q(\varphi(k)\varphi(m)/4)+1$ is prime.*

Remark 4.4. See [11, A239232 and A234475] for related data and graphs. The conjecture implies that there are infinitely many primes of the form $p(n)+1$ (or $p(n)-1$, or $q(n)+1$) with n a positive integer.

Conjecture 4.5. (2013-12-29) *Any integer $n > 7$ can be written as $k+m$ with k and m positive integers such that $p = p_k + \varphi(m)$ and $q(p)-1$ are* *both prime. Also, any integer $n > 7$ not equal to 15 can be written as $k+m$ with k and m positive integers such that $p = p_k + \varphi(m)$ and $q(p)+1$ are both prime.*

Remark 4.5. This implies that there are infinitely many primes p with $q(p)-1$ (or $q(p)+1$) prime. See [11, A234615 and A234644] for related data and graphs. We also conjecture that for any integer $n > 14$ there exists a prime p with $n < p < 2n$ such that $q(p)+1$ is prime.

Conjecture 4.6. (2014-01-07) *For any integer $n \geqslant 60$, there is a positive integer $k < n$ such that $m \pm 1$ and $q(m)+1$ are all prime, where $m = \varphi(k) + \varphi(n-k)/4$.*

Remark 4.6. This implies that there are infinitely many positive integers m with $m \pm 1$ and $q(m)+1$ all prime. We have verified the conjecture for n up to 10^5. See [11, A235343 and A235344] for related data and graphs.

Conjecture 4.7. (2014-01-25) (i) *For any integer $n \geqslant 128$, there is a positive integer $k < n$ such that $r = \varphi(k) + \varphi(n-k)/6 + 1$ and $p(r)+q(r)$ are both prime.*

(ii) *For every $n = 18, 19, \ldots$, there is a positive integer $k < n$ such that $m = \varphi(k)/2 + \varphi(n-k)/8$ is an integer with $p(m)^2 + q(m)^2$ prime.*

Remark 4.7. Clearly, part (i) implies that there are infinitely many primes of the form $p(r)+q(r)$ with r prime. And part (ii) implies that there are infinitely many positive integers m with $p(m)^2 + q(m)^2$ prime. We have verified parts (i) and (ii) for n up to 30000 and 65000 respectively. See [11, A236419, A236412 and A236413] for related data and graphs.

For any positive integer n, $\bar{q}(n) = p(n) - q(n)$ is the number of ways to write n as a sum of unordered positive integers with some part repeated (or even).

Conjecture 4.8. (2014-01-25) (i) *For any integer $n \geqslant 99$, there is a positive integer $k < n$ such that $p = \varphi(k)/2 + \varphi(n-k)/12 + 1$ and $\bar{q}(p)$ are both prime.*

(ii) *For any integer $n > 3$, there is a positive integer $k < n-2$ such that $q(m)^2 + \bar{q}(m)^2$ is prime, where $m = k + \varphi(n-k)/2$.*

Remark 4.8. Clearly, part (i) implies that there are infinitely many primes of the form $\bar{q}(p)$ with p prime. And part (ii) implies that there are infinitely many positive integers m with $q(m)^2 + \bar{q}(m)^2$ prime. See [11, A236417, A236439 and A236440] for related data and graphs.

Conjecture 4.9. (i) (2014-03-12) *Let $n > 1$ be an integer. Then the number $kp(n)q(n)\bar{q}(n) - 1$ is prime for some $k = 1, \ldots, n$. Also, $2p(k)p(n)q(n)\bar{q}(n) + 1$ is prime for some $k = 1, \ldots, n-1$.*

(ii) (2014-01-26) *Any integer $n > 2$ can be written as $k + m$ with k and m positive integers such that $q(k) + \bar{q}(m)$ is prime.*

(iii) (2013-12-08) *For every $n = 2, 3, \ldots$, there is a positive integer $k < n$ with $2^k - 1 + q(n-k)$ prime.*

Remark 4.9. See [11, A239207, A236442 and A233390] for related data and graphs. We have verified the two assertions in part (i) for n up to 83000 and 50000 respectively. We have also checked part (iii) for n up to 2×10^5; for example, for $n = 147650$ we may take $k = 17342$ so that $2^k - 1 + q(n-k)$ is prime.

Conjecture 4.10. (2014-04-24) (i) *For any prime p, there exists a primitive root $g < p$ modulo p which is also a partition number (i.e., $g = p(n)$ for some positive integer n).*

(ii) *For any prime $p > 3$, there exists a primitive root $g < p$ modulo p which is also a strict partition number (i.e., $g = q(n)$ for some positive integer n).*

Remark 4.10. We have verified parts (i) and (ii) for all primes below 2×10^7 and 5×10^6 respectively; see [11, A241504 and A241516] for related data and graphs. We also conjecture that for any prime p there is a primitive root $g < p$ modulo p with $g - 1$ a square (cf. [11, A239957 and A241476]).

References

1. M. Abramowitz and I. A. Stegun (eds.), Handbook of Mathematical Functions with Formulas, Graphs, and Mathematical Tables, 9th printing, New York, Dover, 1972.
2. J.-R. Chen, *On the representation of a large even integer as the sum of a prime and a product of at most two primes*, Sci. Sinica **16**(1973), 157–176.
3. R. Crandall and C. Pomerance, Prime Numbers: A Computational Perspective, Springer, New York, 2001.
4. B. Green and T. Tao, *The primes contain arbitrary long arithmetic progressions*, Annals of Math. **167**(2008), 481–547.
5. G. H. Hardy and S. Ramanujan, *Asymptotic formulae in combinatorial analysis*, Proc. London Math. Soc. **17**(1918), 75–115.
6. D. R. Heath-Brown, *Primes represented by $x^3 + 2y^3$*, Acta Math. **186**(2001), 1–84.
7. H. A. Helfgott, *The ternary Goldbach conjecture is true*, preprint, 2013, arXiv:1312.7748.

8. K. Ireland and M. Rosen, A Classical Introduction to Modern Number Theory, 2nd Edition, Graduate texts in math., vol. 84, Springer, New York, 1990.
9. P. Ribenboim, The Little Book of Bigger Primes, 2nd Edition, Springer, New York, 2004.
10. L. Schoenfeld, *Sharper bounds for the Chebyshev functions $\theta(x)$ and $\psi(x)$. II*, Math. Comp. **30**(1976), 337–360.
11. Z.-W. Sun, Sequences A182662, A185636, A210210, A218829, A230022, A231201, A232443, A232463, A232465, A232502, A232861, A233150, A233183, A233206, A233296, A233390, A233439, A233529, A233544, A234470, A234475, A234567, A234569, A234615, A234644, A234694, A234695, A235061, A235189, A235343, A235344, A235924, A235925, A235934, A235935, A235984, A236066, A236097, A236119, A236193, A236241, A236242, A236412, A236413, A236417, A236419, A236439, A236440, A236442, A236481, A236548, A236566, A236531, A236831, A236832, A237259, A237260, A237284, A237291, A237413, A237414, A237453, A237497, A237531, A237578, A237582, A237595, A237597, A237598, A237612, A237615, A237643, A237657, A237687, A237706, A237710, A237712, A237768, A237769, A237815, A237817, A237837, A237840, A237879, A237975, A238134, A238224, A238277, A238278, A238281, A238289, A238457, A238458, A238459, A238504, A238509, A238570, A238573, A238576, A238585, A238643, A238645, A238646, A238701, A238703, A238732, A238766, A238814, A238878, A238881, A238890, A238902, A239207, A239209, A239214, A239232, A239328, A239330, A239428, A239430, A239451, A239884, A239957, A240604, A241476, A241504, A241516, A242425, A242748, A242753, A242950, A243164, A243403 in OEIS (On-Line Encyclopedia of Integer Sequences), `http://oeis.org`.
12. Z.-W. Sun, *On functions taking only prime values*, J. Number Theory **133**(2013), 2794-2812.
13. Z.-W. Sun, *Conjectures involving arithmetical sequences*, in: Number Theory: Arithmetic in Shangri-La (eds., S. Kanemitsu, H. Li and J. Liu), Proc. 6th China-Japan Seminar (Shanghai, August 15-17, 2011), World Sci., Singapore, 2013, pp. 244-258.
14. Y. Zhang, *Bounded gaps between primes*, Annals of Math. **179**(2014), 1121–1174.

INDEX

H-function, 13
K-Bessel function, 18
L-function of elliptic curve, 102
X-function, 17
Y-function, 32
j-invariant, 157
n-th prime, 169
σ-series, 39

abelian extension, 79
Abelian variety, 145, 163
almost periodicity, 117
arithmetic progression, 172, 173
Artin L-function, 102
Atkinson disscetion, 62
automorphic L-function, 101
automorphism, 79

Barnes multiple zeta-function, 114
Bertrand's postulate, 172
beta transform, 22

Catalan's conjecture, 1
catalytic Dirichlet series, 63
Chebotarev's density theorem, 75
class number, 123
CM field, 164
combinatorial properties, 170
complete system of residues modulo n, 170, 174
completely splitting, 75
complex multiplication, 149, 154, 164
composite universality, 130
confluent hypergeometric function, 22
congruence condition, 79
conjectures, 170

decomposable, 77
decomposable decomposition, 77
Dedekind zeta-function, 15, 100
denseness theorem, 97
density, 78
Dirichlet L-function, 100
discontinuous integral, 52
discrete universality, 120
discriminant, 24
distribution of roots, 75
divisor sum, 67

elliptic curve, 153, 154
elliptic function, 149
elliptic function field, 150, 156
elliptic modular function, 157
endomorphism algebra, 154, 164
Epstein zeta-function, 128
ergodic universality, 132
Estermann zeta-function, 109
Euler's totient function, 173
Euler-Zagier sum, 114
Eulerian number, 77

Fermat primes, 1
figurate prime, 2
first generalized Katsurada formula, 53
functional equation, 15, 16, 29, 39
Functional limit theorem, 98

gamma transform, 20
generalization of Wilton's Riesz sum, 60
generalized Bessel function, 23

Goldbach's conjecture, 3, 175, 181
Green-Tao theorem, 178

Hecke L-function, 102
Hecke type functional equation, 34
Hilbert class field, 159
Hurwitz zeta-function, 107
hybrid universality, 127
hyperelliptic curve, 161

ideal class group, 159
Igusa zeta-function, 128
improper modular relation, 52
indecomposable, 77
inverse Heaviside integral, 21
inverse of k mod p_k, 183

Jacobian variety, 162
joint universality, 104

key function, 14
Koshlyakov's K-function, 39

Legendre's conjecture, 174
Lerch zeta-function, 109
limit theorem, 98

Maass form L-function, 103
Matsumoto zeta-function, 101
Meijer G-function, 13
Mesernne primes, 1
mixed universality, 115
modular relation, 14
module of periods, 148, 150

number field, 15

Oppenheim-Wilton's formula, 61
order, 158
ordering, 76

partition function, 183
partition number, 186
period matrix, 163
periodic Hurwitz zeta-function, 109
periodic zeta-function, 101
Perron's formula, 52
positive density method, 101
Prime Number Theorem, 169
primes, 169
primitive root, 183, 186
processing gamma factor, 13
proper figurate prime, 5

quotient of gamma factors, 39

Ramanujan's formula, 37
Rankin-Selberg L-function, 102
reducible, 76
reduction-augmentation formula, 19
residual function, 17, 26
Riemann Hypothesis, 3
Riemann hypothesis, 117
Riemann relations, 163
Riemann type functional equation, 31
Riemann's Hypothesis, 169
Riesz mean, 50
Riesz sum, 23, 50, 52
Riesz summable, 50

sampling theorem, 36
second generalized Katsurada formula, 57
Selberg class, 102
Selberg zeta-function, 103
shifts universality principle, 114
singular modulus, 158
sinus cardinalis function, 35
Sophie Germain prime, 175
spectral zeta-function, 128
square, 176
squarefree, 176
Steen's function, 24
Steuding class, 102
strict partition function, 183
strict partition number, 186
strong recurrence, 117
strong universality, 108
subject, 95–97
super twin prime, 177

Super Twin Prime Conjecture, 177
symmetric power L-function, 102

theta-transformation formula, 28
triangular number, 176
truncated Riesz sum, 52
twin prime, 171, 173, 175, 176, 178, 181, 182
twin prime conjecture, 177, 178
twin primes conjecture, 2

uniform distribution, 78
unprocessed modular relation, 27

Voronoǐ formula, 62
Voronoǐ function, 23

weak Goldbach conjecture, 181
Weierstrass $\wp$ function, 151
Weierstrass' equation, 151
weighted universality, 119
Whittaker function, 22
Wilton's approximate functional equation, 66
Wilton's generalized Bessel function, 60

zeta-function of symmetric matrices, 128